Rollover Accident Reconstruction

R-475 Related Resources

Vehicle Accident Analysis and Reconstruction Methods
Author: Raymond M. Brach, Matthew Brach
PRODUCT CODE: R-397

Integrated Automotive Safety Handbook
Author: Mark Gonter, Ulrich W. Seiffert
PRODUCT CODE: R-407

Crash Reconstruction Research
Author: Michael S. Varat
PRODUCT CODE: PT-138

Commercial Vehicle Accident Reconstruction and Investigation
Author: Rick W. Varner, Roy F. Sutphen
PRODUCT CODE: B-883

Accident Reconstruction: Crash Analysis
Author: Raymond M. Brach, Stein E. Husher, Alan W. Thebert, Michael S. Varat
PRODUCT CODE: SP-1572

Accident Reconstruction: Analysis, Simulation, and Visualization
Author: Wesley D. Grimes, Terry D. Day, Raymond M. Brach, Alan W. Thebert
PRODUCT CODE: SP-1491

Rollover Accident Reconstruction

NATHAN ROSE
Principal and Director
Kineticorp, LLC
nrose@kineticorp.com

GRAY BEAUCHAMP
Principal Engineer
Kineticorp, LLC
gbeauchamp@kineticorp.com

ALAN F. ASAY
Asay Engineering, LLC
alan@asayengineering.com

SAE INTERNATIONAL®

Warrendale, Pennsylvania, USA

400 Commonwealth Drive
Warrendale, PA 15096-0001 USA
E-mail: CustomerService@sae.org
Phone: 877-606-7323 (inside USA and Canada)
724-776-4970 (outside USA)

Library of Congress Catalog Number 2018935384
SAE Order Number R-475
http://dx.doi.org/10.4271/R-475

ISBN-Print 978-0-7680-9372-8
ISBN-HTML 978-0-7680-9373-5
ISBN-epub 978-0-7680-9374-2
ISBN-prc 978-0-7680-9375-9

To purchase bulk quantities, please contact: SAE Customer Service

E-mail: CustomerService@sae.org
Phone: 877-606-7323 *(inside USA and Canada)*
724-776-4970 *(outside USA)*

Visit the SAE International Bookstore at books.sae.org

dedication

Nathan dedicates this book to Susie, Alex, Brooklyn, Hudson, Morgan, Anna, Adriana, and Addison. I love doing life with you guys!

contents

preface

I first taught the Society of Automotive Engineers' *Reconstruction and Analysis of Rollover Crashes of Light Vehicles* (Course C1502) in April of 2015, a task I was initially hesitant to take on because of the time commitment involved. Once I made the decision to teach the course, though, and began preparing the materials, I started to see my role as that of curator, and I more clearly saw the benefit I would get from the time investment. The science of reconstructing rollover crashes has been a work in progress for decades. Particularly in the last 10 to 15 years, though, the science has progressed significantly and, it seemed to me, it would be useful to pull together in one place much of what has been learned. I knew that, in my own practice as an accident reconstructionist, I would use a volume that accomplished that. That said, reviewing the extensive literature related to reconstructing rollover crashes and pulling it together into a coherent volume has indeed been a time-consuming task. I began preparing this book (which started as my course materials) sometime in late 2014, and it has been evolving and growing up to the present.

The first version of my course notes covered models and results from my own papers in far too much detail - the glassy-eyed looks from the first students who took my course told me that. These models tended to be heavy on the underlying mathematics. On top of that, all the results were covered in English units alone. I know, I know, but an important part of my job involves sharing my findings with juries in the United States. These juries just cannot think in kilometers per hour and meters (and, of course, neither can I). I have since cut out significant portions where the coverage of models was needlessly mathematical and theoretical. I have tried to make this book more practical than when it started out. I have also added metric units to the text and added a metric version of all of the graphs. For now, these graphs are consigned to the outer darkness, in appendices. If you have a suggestion for how I can more fluidly incorporate them, I am certainly open to your suggestion. Email me at nrose@kineticorp.com. I am also happy to hear your suggestions and comments about what portions of this text still make your eyes glaze over and about what important pieces of literature I still have not cited and summarized.

Finally, I have added Gray Beauchamp and Alan Asay as collaborators on this project. Each of them brings extensive expertise and experience analyzing rollover crashes, and each of them has made substantial contributions to the science of rollover reconstruction. In some of what follows, we have broken out of the role of curator and have presented new analysis. At one point, for instance, we realized that no one had gone through and tested the most widely used speed analysis methods against the extensive library of heavily documented, full-scale, steering-induced rollover tests that now exist in the literature. We have now completed that analysis and Chapter 8 covers much of what we learned.

Our understanding of how to reconstruct rollovers continues to evolve, and I'm sure there will eventually be a second edition of this book. I envision adding a section on electronic stability control and several rollover case studies involving vehicles from which event data was imaged. In the meantime, we hope you enjoy this first edition!

Nathan Rose
Principal and Director
Kineticorp, LLC
March, 2017

acknowledgments

We would like to acknowledge and thank the following members of Kineticorp's technical staff who assisted with preparation of graphics, illustrations, and vehicle data for this book: Neal Carter, Jordan Dickinson, Alireza Hashemian, Andrew O'Donnell, Martin Randolph, and Toby Terpstra. Thanks, also, to Jarrod Carter for reviewing and commenting on parts of this book and to Rob Larson and Exponent for allowing us to include frames from several of their crash test videos.

CHAPTER 1

Overview of Rollover Reconstruction

According to the National Highway Traffic Safety Administration (NHTSA), "of the nearly 9.1 million passenger car, SUV, pickup and van crashes in 2010, only 2.1% involved a rollover. However, rollovers accounted for nearly 35% of all deaths from passenger vehicle crashes. In 2010 alone, more than 7,600 people died in rollover crashes. The majority of them (69%) were not wearing safety belts."[1,2] According to the Insurance Institute for Highway Safety, approximately 7,200 people died in rollover crashes in 2015 (32% of fatalities).[3] These statistics include all rollover crash types - single-vehicle crashes and rollover crashes preceded by an impact with another vehicle. According to data reported by Hughes [1], 85% of rollover crashes involve the vehicle rolling 1 complete revolution or less. Only 2.3% involve the vehicle rolling more than 2 complete revolutions.

Rollover crashes can be divided into a number of categories based on their attributes. Some rollovers involve a single vehicle; others occur following a vehicle-to-vehicle impact and, thus, involve different forces and mechanisms than single-vehicle rollovers. Some rollovers occur on a paved roadway and others occur off the roadway

[1] https://www.safercar.gov/Vehicle-Shoppers/Rollover/Fatalities

[2] There were a total of 32,885 crash fatalities in the US in 2010. For 2011 through 2015, total crash fatalities in the US varied between 32,367 and 35,092. See "2010 Motor Vehicle Crashes: Overview," US Department of Transportation, National Highway Traffic Safety Administration, DOT HS 811 552, February 2012. See also similarly titled documents for 2011 through 2015.

[3] http://www.iihs.org/iihs/topics/t/general-statistics/fatalityfacts/passenger-vehicles/2015

on dirt, gravel, and vegetation. Some rollovers terminate when all of the vehicle energy is dissipated through vehicle-to-ground impacts; others terminate when the vehicle strikes a fixed object. Some rollovers are initiated through a combination of severe steering inputs and suspension effects, others are initiated when the vehicle strikes a curb or furrows in soil, and some rollovers are caused by or preceded by impacts.

An informational report issued by the Society of Automotive Engineers (SAE) titled "Rollover Testing Methods" lists the following terminology relevant to defining the characteristics of a rollover, particularly single-vehicle rollovers (2011). These terms distinguish the rollovers in terms of the nature of the forces that initiate the rollover. In these definitions below, the wording from the original SAE document has been updated and refined and, for some of them, we have also offered commentary.

On-road rollover - A rollover that is initiated on a paved road surface.

Untripped rollover - A rollover that is initiated on the road surface due to friction forces between the tires and the road surface. This term can be misleading and this book does not use this term. The issue with the use of this term is that the forces generated between a passenger vehicle's tires and a paved roadway are lateral tripping forces. While these tripping forces are typically insufficient to cause a rollover absent additional contributing factors, such as aggressive steering inputs with suspension effects, they nonetheless can trip the vehicle when combined with other conditions. It seems more meaningful to think of tripping forces along a spectrum with tire-roadway frictional forces typically being the lowest in magnitude, soil frictional and furrowing forces being next, and curb tripping forces being the greatest in magnitude.

Tripped rollover - A rollover that is initiated when the lateral motion of the vehicle is slowed or stopped quickly enough to generate the rolling motion. The opposing force may be produced by vehicle tires and wheels furrowing into soil (a soil-tripped rollover), impacting a raised curb (a curb-tripped rollover), or impacting some other object.

Pitch-over - A rollover in which the vehicle rotates end over end through an angle greater than 90°.

Turn-over - The SAE informational report defines a turn-over as a rollover that is initiated when centrifugal forces from a sharp turn or vehicle rotation are resisted by the surface friction without furrowing or gouging. However, the forces that can be generated between a passenger vehicle's tires and a paved roadway are typically insufficient to cause a rollover, absent some other factor. So, these rollovers would only occur with some other factor also present - perhaps a shift in the center of gravity due to modifications or cargo placement on or in the vehicle.

Fall-over - A rollover that is initiated when the surface on which the vehicle is traveling slopes downward in the direction of vehicle movement so that the center of gravity moves outboard of the leading tires.

Bounce-over - When a vehicle strikes a fixed object, rebounds, and overturns. The rollover usually occurs near the impacted object.

Flip-over - A rollover initiated by a vehicle rotating around its longitudinal axis after engaging a ramp-like object, such as a turned-down guardrail or the back slope of a ditch. The vehicle may be in a yaw and sideslipping when it contacts the ramp-like object. In a laboratory setting, this type of rollover is initiated with an actual ramp and is referred to as a corkscrew rollover.

This book focuses on tripped, single-vehicle rollover crashes that terminate without striking a fixed object. Figure 1.1 depicts a typical scenario that will be addressed. This scenario involves the following. (1) A driver inputs a rightward steer of sufficient severity to cause the vehicle to enter a clockwise yaw with sideslip; oftentimes, the steering input that leads to a loss of control is preceded by other steering inputs. (2) The driver may input a countersteer back to the left, but the steer is insufficient to reverse the vehicle's yaw motion and sideslipping. (3) At some point during the clockwise yaw, the trailing-side[4] tires lift off the roadway; in the example here, the liftoff of the tires is due to the leading-side tires furrowing into the soil. (4) As this motion continues, the center of mass travels over the leading-side tires, the leading-side[5] tires lift off the roadway, and the vehicle becomes airborne and begins rolling. (5) The vehicle rolls until its kinetic energy is dissipated and it comes to rest.

A rollover crash like the one depicted in Figure 1.1 is often split into three phases for analysis - the loss-of-control phase or yaw phase, the trip phase, and the roll phase. During the loss-of-control phase, driver steering inputs - or perhaps some other external force - result in the driver losing control of the vehicle. The vehicle enters an unrecoverable yaw during this phase. Analysis of this phase is often aimed at determining the speed of the vehicle when the driver lost control; determining the steering, braking, or acceleration inputs that led to the loss of control; and assessing the reasonableness of those inputs.

The trip phase begins when the trailing-side tires of the vehicle (the right-side or passenger-side tires in the example of Figure 1.1) lift off the ground and ends when the leading-side tires also lift off and the vehicle becomes airborne. Analysis of this phase is often aimed at determining the speed of the vehicle at the beginning of trip, determining the mechanism that caused the trip (i.e., soil, curb, etc.), determining the forces applied to the vehicle by that mechanism, calculating the roll velocity and vertical velocity with which the vehicle entered the roll phase, and analyzing how the occupants moved during this phase.

FIGURE 1.1 Typical high-speed, single-vehicle rollover crash.

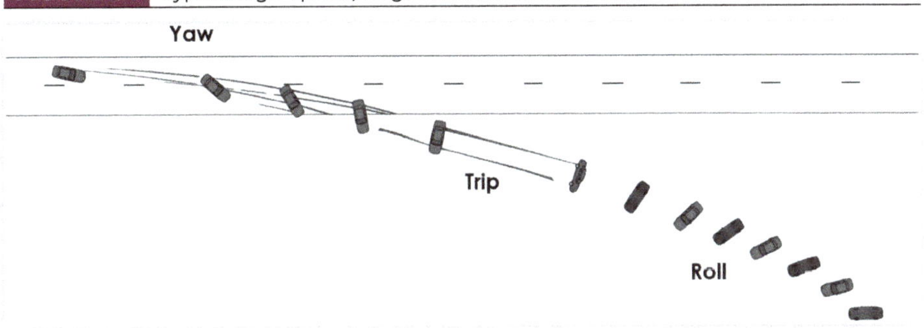

[4] Sometimes referred to as the far side.

[5] Sometimes referred to as the near side.

The roll phase begins when the vehicle becomes airborne. In a rollover that occurs at highway speeds, the vehicle will typically experience multiple vehicle-to-ground impacts during the roll phase. The roll velocity of the vehicle generally ramps up rapidly during the early portions of the roll phase and may level off temporarily after reaching a peak. Beyond the peak, the roll rate tends to ramp down slower than it ramped up until the vehicle comes to rest. Analysis of the roll phase is often aimed at determining the speed of the vehicle when it began rolling, determining the number of rolls the vehicle experienced, assessing the variation of the roll velocity, identifying ground impact locations (along with determining the forces applied to the vehicle during those ground impacts), determining when a particular component was damaged, determining if the occupants were belted, and analyzing when and where during the roll occupants were ejected or injured.

This book addresses each of the phases of a rollover crash in turn. The order flows in the order that a typical reconstruction flows - from the roll phase, to the trip phase, to the yaw phase. Clearly, not all rollovers involve all of these phases and, for any particular reconstruction, the analyst will have to determine which models are most relevant.

The remainder of this chapter introduces a general approach to reconstructing rollover crashes and gives an overview of the test procedures and data that are available in the literature for analyzing this type of crash. The second chapter introduces the types of physical evidence that are commonly deposited at the accident scene and on the vehicle during a rollover crash, and discusses methods for documenting that evidence. Chapters 3-6 address each phase of a typical rollover crash in detail. Chapter 3 introduces analysis of the roll phase and describes a basic approach. Chapter 4 introduces more advanced analysis methods for the roll phase. Chapter 5 addresses the trip phase and Chapter 6 addresses the yaw phase. Chapter 7 describes analysis methods for tire mark striations. Chapter 8 utilizes the analysis techniques described in Chapters 3-6 to analyze all of the full-scale, steering-induced loss-of-control rollover crash tests available in the literature and reports error rates for the speed analysis techniques. Chapter 9 describes analysis of rollover crashes with simulation. Chapter 10 describes the use of data from event data recorders (EDRs) in rollover crash reconstruction. Chapter 11 describes methods for analyzing occupant ejections in rollover crashes.

General Speed Analysis Approach

During a high-speed, single-vehicle rollover crash, the vehicle's initial kinetic energy is dissipated over the course of the yaw, the trip, and the roll phases. This can be described mathematically with the following energy balance, where the energy loss during each phase of the crash is calculated based on the distance traveled during each phase (d_{yaw}, d_{trip}, and d_{roll}) and the corresponding deceleration rates or drag factors (f_{yaw}, f_{trip}, and f_{roll}). In this equation, m is the vehicle mass, W is the vehicle weight, and v_i is the vehicle's initial velocity. Typically, the initial velocity is calculated at the beginning of visible tire marks deposited from the loss of control[6]:

$$\frac{1}{2}mv_i^2 = f_{yaw}Wd_{yaw} + f_{trip}Wd_{trip} + f_{roll}Wd_{roll} \tag{1.1}$$

[6] There is often some speed loss prior to the beginning of the tire marks. In some cases, this is due to multiple steering inputs that preceded the loss of control. Oftentimes, the only evidence of these additional steering inputs will be witness or driver statements or testimony. Speed loss from these prior steering inputs could potentially be quantified with simulation.

Equation (1.1) can be rewritten as follows to yield the speed at the beginning of the loss of control:

$$v_i = \sqrt{2g\left(f_{yaw}d_{yaw} + f_{trip}d_{trip} + f_{roll}d_{roll}\right)} \tag{1.2}$$

Similar expressions could be written to yield the speeds at the beginning of the trip and roll phases. Equations (1.1) and (1.2) assume that the initial rotational kinetic energy (yaw) is negligible relative to the translational energy. This is typically a reasonable assumption for single-vehicle rollovers. Rose [2] covers methods that do not make this assumption. These methods could be applicable to a rollover preceded by an impact that induces a high yaw velocity.

Reconstructing a rollover crash generally includes the following steps:

- Documenting, mapping, and diagraming the geometry of the accident site (slope, cross-slope, lane widths, shoulder width, roadway surface type, off-road surface type, etc.). This often involves physically visiting the accident site in order to obtain the necessary measurements.
- Documenting, mapping, and diagraming the physical evidence deposited at the scene. Some of this evidence may still be present during an accident site inspection. Other evidence locations may need to be reconstructed based on measurements taken by police or using methods of photogrammetry in conjunction with scene photographs.
- Documenting, mapping, and diagramming the precrash geometry of the vehicle. At times, this involves obtaining manufacturer specifications for the vehicle. At other times, it involves inspecting an exemplar vehicle or performing calculations related to the inertial properties of the vehicle.
- Documenting and diagraming the physical evidence deposited on the vehicle. This often involves physically inspecting the vehicle, but could also rely on photogrammetric analysis.
- Determining the vehicle motion that accounts for the physical evidence identified at the scene and on the vehicle.
- Determining the piecewise average deceleration rates that this motion implies and quantifying the vehicle speed at a number of points during the sequence.
- A reconstruction may also involve a more detailed timing analysis in order to make more specific determinations, such as the roll velocities experienced during the roll phase, the location, timing, and severity of specific vehicle-to-ground impacts, the location at which certain windows broke, or the location when specific portions of the vehicle were damaged.

Vehicle Geometry and Inertial Parameters

Vehicle dimensions often need to be obtained for reconstructing a rollover crash. At a minimum, these dimensions will include the length, width, and height of the vehicle, along with the wheel locations. These dimensions can be obtained from manufacturer specifications or from inspecting and measuring an exemplar vehicle. Depending on the analysis techniques employed, the reconstructionist may also need to determine the weight of the vehicle, locate its center of mass, and estimate its moments of inertia.

The moments of inertia characterize the vehicle's resistance to rotation about its principal axes - roll, pitch, and yaw.

A basic estimate of the vehicle moments of inertia can be obtained with the prism method, which assumes that the vehicle is a solid, homogeneous box. With this assumption, the moments of inertia about the principal axis are given by the following equations:

$$I_{roll} = \frac{1}{12}m\left(w_{overall}^2 + h_{overall}^2\right) \tag{1.3}$$

$$I_{pitch} = \frac{1}{12}m\left(l_{overall}^2 + h_{overall}^2\right) \tag{1.4}$$

$$I_{yaw} = \frac{1}{12}m\left(l_{overall}^2 + w_{overall}^2\right) \tag{1.5}$$

In these equations, I_{roll}, I_{pitch}, and I_{yaw} are the moments of inertia about the roll, pitch, and yaw axes. The vehicle mass is indicated with the letter m, and $l_{overall}$, $w_{overall}$, and $h_{overall}$, are the vehicle length, width, and height, respectively.

MacInnis [3] examined this and other methods for estimating the whole-vehicle (as opposed to sprung mass) moments of inertia. He compared each method to actual moments of inertia measured and reported by the NHTSA [4, 5, 6]. In making this comparison, MacInnis divided the vehicles into the following categories: front-wheel-drive passenger cars, rear-wheel-drive passenger cars, sport utility vehicles (SUVs), pickup trucks, and vans. The article by MacInnis gives a complete listing of the equations he recommended for calculating the center of gravity height and the moments of inertia for each of these vehicle types. Additionally, MacInnis examined the effect of uncertainty in the yaw moment of inertia on the results of planar collision simulation with the accident reconstruction software PC-Crash. He found that varying the yaw moment of inertia by 30% had about a 3% effect on the calculated initial speeds.

The NHTSA published additional moments of inertia data after the MacInnis study [7]. Allen [8] incorporated this additional data and used regression analysis to develop equations for estimating the vehicle center of mass height and moments of inertia. Allen did not partition the data by vehicle type. For each moment of inertia, Allen's equations took the following form:

$$I_i = 10^{k1} \cdot l_{wb}^{k2} \cdot T^{k3} \cdot h_{overall}^{k4} \cdot W^{k5} \tag{1.6}$$

In these equations

I_i is whichever moment of inertia is under consideration
l_{wb} is the wheelbase (in feet)
T is the average track width (in feet)
$h_{overall}$ is the vehicle height (in feet)
W is the total vehicle weight (in pounds)

The exponents in Equation (1.6) are the regression coefficients. The values of these coefficients for each of the moments of inertia are reported in Table 1.1. With the regression coefficients of Table 1.1, Equation (1.6) yields the moments of inertia in units of pounds-feet-sec^2.

TABLE 1.1 Regression coefficients from Allen [8]

	k1	*k2*	*k3*	*k4*	*k5*
I_{roll}	−2.1363	−0.1596	1.9404	0.3629	0.9421
I_{pitch}	−2.0024	1.5315	0.2526	0.1009	1.0206
I_{yaw}	−1.7797	1.4316	0.3811	0.0188	0.9800

Allen also reported the following equation for estimating the vehicle center of gravity height (in feet). Allen reported an R^2 value for this equation of 0.8277:

$$h = 0.3891 \cdot h_{overall} + 0.113 \tag{1.7}$$

Garrott [9] tested the effect of adding occupants and cargo on the center of gravity height of a number of vehicles. In nearly every case, he found that the center of gravity height increased with the addition of occupants. The effect of adding cargo to the vehicle varied depending on how the cargo was placed. Some cases may warrant physical measurement of a particular vehicle's center of mass height, either in the empty or loaded condition. Shapiro [10] discussed the pros and cons of four different methods for making a physical measurement of a vehicle's center of mass height.

Overview of Rollover Test Procedures

This book utilizes rollover crash test data to develop methods for analyzing each of the phases of a rollover crash. As a precursor to that, this section introduces test procedures that have been used by researchers to study rollover crashes and summarizes some of the results from these studies. The focus here is on full-vehicle testing where the vehicle rolls freely after the roll has been generated, although the literature also contains component-level test procedures (see Achhammer [11], for example; SAE J2926 describes others, such as the Jordan Rollover System test).

Wilson and Gannon [12] summarized the history and state of rollover testing in 1972. They noted that the earliest recorded rollover test with automobiles was run at the General Motors Proving Ground in 1934. "The rollover test at that time consisted of tipping a car on its side and allowing it to roll over, down a hill…Other early rollover tests consisted of driving a car into a spiral ramp located at the top of a hill and inducing ground level rollovers by driving the car into a skid on a sod field. Other rollover tests in the 1930's were conducted by Chrysler. Their tests consisted of an end-over-end roll produced by pushing a vehicle off a steep cliff at Bald Mountain, northwest of Pontiac, Michigan." Wilson and Gannon noted that these early efforts failed to produce test procedures that were repeatable and mimicked real-world crashes.

One solution to the problem of repeatability was the development of laboratory tests for vehicle roof strength. Wilson and Gannon discuss several such test procedures, including an inverted drop test, a test in which a flat impact plate weighing as much as the vehicle was dropped onto the roof, and another test in which a flat rigid plate was pressed into the roof structure. While these procedures increased the repeatability of the testing, they did not mimic real-world crashes and, as Wilson and Gannon note, field data showed "no significant correlation between increased roof crush and increased head injury or increased overall occupant injury." The first repeatable (at least in terms of the roll initiation) and realistic test procedure was the dolly rollover test originally developed and presented by Mercedes-Benz in 1970 and then incorporated into Federal Motor Vehicles Safety Standard 208 in 1971. Through the years, the industry has continued to develop this test procedure, and it has become one of the staples of rollover crash testing.

FIGURE 1.2 Dolly rollover test setup.

Dolly Rollover Tests

A dolly rollover test involves generating a lateral roll of the test vehicle by accelerating the vehicle on a cart up to the test speed (Figure 1.2). In the standard dolly rollover test, the vehicle is situated at an initial roll angle of 23° on a cart with the leading-side tires resting against a 4-in. (10.16 cm) flange that helps to initiate the rollover when the cart is quickly decelerated from a speed of approximately 30 mph. The NHTSA has used this test procedure to test numerous passenger vehicles [13]. The SAE Recommended Practice describing the standard dolly rollover procedure was adopted in 1993 [14]. This dolly rollover test procedure has been used to evaluate restraint system performance, rollover occupant protection system performance, occupant kinematics and ejection, vehicle structural integrity, and vehicle rollover kinematics.

Orlowski et al. [15, 16] reported eight dolly rollover tests conducted with 1983 Chevrolet Malibus at speeds of approximately 32 mph (51.5 km/h).[7] These Malibus were front-engine, rear-wheel-drive cars that weighed approximately 3,200 lb. Four of the vehicles had standard production roofs and four had modified roofs with roll cages. In these tests, the vehicles were launched laterally from the dolly onto a flat and dry asphaltic concrete surface with their right sides leading and with an initial roll angle of 23°. Unrestrained 50th percentile Hybrid III dummies were situated in the driver's and passenger's side front seats. In these tests, the vehicles with standard roofs rolled between 2½ and 3½ revolutions over distances between 66 and 89 ft (20.1 to 27.4 m). The vehicles with roll cages rolled between 2 and 3½ revolutions over distances between 65 and 74 ft (19.8 and 22.6 m).

In 2007, Luepke [18] reported two dolly rollover tests, one utilizing a 1998 Ford Expedition and the other utilizing a 2004 Volvo XC90. In these two tests, the vehicles were launched onto a 6-in. (15 cm) deep compacted desert-type dirt surface. At the beginning of both tests, the vehicle was positioned on the dolly with the driver's side leading and with an initial roll angle of 23°. The leading-side tires were positioned against a 4-in. (10.2 cm) flange with the leading tires initially 9 in. (23 cm) above the ground surface. Prior to the vehicles exiting the dolly in these tests, the dolly was accelerated up to a nominal speed of 43 mph (70 kph) [Expedition - 43.2 mph (69.7 kph); XC90 - 42.9 mph (69.2 kph)]. The dolly was quickly decelerated over a distance of 5-6 ft (1.5 to 1.8 m) to initiate the roll of the vehicle. The Ford Expedition rolled 4 times in approximately 120 ft (37 m). The Volvo XC90 completed 4¼ rolls in approximately 115 ft (35 m).

In 2008, Luepke [19] reported three additional dolly rollover tests on a desert-type earth surface utilizing the following test vehicles: a 1997 Ford Aerostar, a 1989 Chevrolet S10 Blazer, and a 1987 Jeep Grand Wagoneer. These three tests had the same setup as the Ford Expedition and Volvo XC90 tests. Just prior to the quick deceleration that initiated the rollovers, the dolly was accelerated to the following speeds: 41.7 mph (67.3 kph) for the Ford Aerostar, 55.3 mph (89.2 kph) for the Chevrolet Blazer, and 40.6 mph (65.5 kph) for the Jeep Grand Wagoneer. The Ford Aerostar rolled 3½ times in 95 ft (29 m). The Chevrolet Blazer rolled 8 times in 204 ft (62.2 m). The Jeep Grand Wagoneer rolled 3¾ times in 105 ft (32 m).

In 2008, Rose [20] reported analysis of a dolly rollover crash test that utilized a Ford SUV. The test vehicle was situated on the cart with an initial roll angle of 23°, accelerated up to the test speed, then decelerated quickly to initiate the rollover. In this particular test, the vehicle was situated on the cart with its driver's side leading and the rollover was initiated from a speed of approximately 31 mph (50 kph). After exiting the dolly, the vehicle's driver's side wheels were the first to contact the ground, and at the time this occurred, the vehicle was traveling approximately 29.5 mph (47.6 kph). Figure 1.3 shows the roll dynamics that occurred during this test. As these images show, the vehicle rolled

[7] See also [17].

FIGURE 1.3 Rollover crash test dynamics from Rose [21].

once and came to rest on its wheels. Prior to the test, the overall geometry of the vehicle was surveyed, as was the geometry of the crash test facility. After the test, additional surveys of the vehicle and the test facility were conducted with the purpose of documenting the posttest condition of the vehicle and the test surface.

In 2010, Croteau [22] reported three passenger-side-leading dolly rollover tests of 2003 Subaru Foresters on concrete and dirt. Two of the Foresters rolled on a dirt surface and one of them rolled on a concrete surface. The release speed of the vehicle from the dolly in Test #1 was 44.2 mph (71.3 kph), in Test #2 it was 44.0 (71.0 kph) mph, and in Test #3 it was 43.4 mph (70 kph). In Test #1, the vehicle completed 5½ rolls in 129 ft (39.3 m). In Test #2, the vehicle rolled twice in 160 ft (48.8 m). In Test #3, the vehicle completed 5¾ rolls in 138 ft (42.1 m).

In 2015, Turriff [23] reported results from four dolly rollover test runs with 2001 Chevrolet Trackers. Each test was run with the driver's side of the vehicle leading and the dolly traveling a nominal speed of 35 mph (56 kph) prior to the sudden deceleration that initiated the separation of the vehicle from the dolly and the rollover. All four vehicles rolled between 1 and 1½ times. The vehicles began rolling on a smooth concrete surface and then traveled onto a level dirt surface where they continued to roll. A small amount of the dirt (a sandy aggregate mixture) was present on the concrete surface. Turriff examined the characteristics of scratches on the vehicles after the tests.

Several commentators have observed that dolly rollover tests initiate a rollover in a manner that is different from real-world rollovers. Many real-world rollovers are preceded by a loss of control that involves steering inputs by the driver, suspension and tire effects, or an impact. Some researchers have hypothesized that differences in the manner of roll initiation could result in differences in the relationships that exist between the following variables: roll initiation speed (trip speed), roll distance, number of rolls, and average deceleration rate [21].

Rollover Tests Using Modified Dolly Fixtures

A number of researchers have conducted rollover crash tests that utilized dolly fixtures that were different than the standard one used for the tests described in the previous section. For example, in 1990, Cooperrider [24] conducted a series of rollover tests that included both curb-tripped and soil-tripped rollovers. These tests used modified dollies, where the test vehicle did not have an initial roll angle but was accelerated up to speed before encountering a curb or resistance from the soil. Out of five attempts, two of the curb trip tests resulted in rollover. Both soil trip tests resulted in rollover. The curb trip tests utilized a 6-in. (15 cm) square section of steel box tubing that was rigidly attached to the ground perpendicular to the vehicle's travel direction. The test vehicles traveled laterally into the curb with their right sides leading at speeds around 30 mph (48 kph). The soil-tripped tests utilized a 40 ft by 20 ft (12.2 by 6.1 m) area of loosely packed dirt and finely crushed granite. Similar to the curb-tripped tests, the test vehicles traveled laterally into this soil with their passenger sides leading. The initial speeds of the vehicles in these two tests were 33.7 and 27.0 mph (54.2 and 43.5 kph). The vehicle tires furrowed into the soil, causing the vehicles to begin to trip and then roll over.

In 1998, Cooperrider [25] reported six additional soil trip tests that each utilized a 1984 Oldsmobile Cutlass Ciera. The motion from one of these tests is illustrated in Figure 1.4. The first test was run with an initial speed of 13.5 mph (21.7 kph) and "the subsequent test speeds were established based on the previous test results in an effort to closely bracket the minimum trip speed." Ultimately, the initial speeds for the vehicles in this test series ranged from 13.5 to 42.9 mph (21.7 to 69.0 kph). In the first four tests, the vehicle did not roll over. The initial speed of the test vehicle in the fourth test was 21.2 mph (34.1 kph). In the fifth and sixth tests, which had speeds of 23.0 mph and 42.9 mph (37.0 and 69.0 kph), the vehicle rolled over - ½ roll in the fifth test and 1 roll in the sixth test.

In 2000, Larson [26] discussed a test procedure that utilized a modified dolly referred to as the Roller Coaster Dolly (Figure 1.5). This test fixture would ride on a pair of tubular rails "in a manner similar to a roller coaster." This fixture allowed a moving vehicle to be released onto a flat or sloped terrain with a preselected roll, pitch, and yaw angle. In explaining the motivation for developing this test procedure, Larson noted that "the majority of rollover accidents occur off-road, where the surface is often sloped or uneven. The inclination of the surface where the rollover occurs can be an important factor in why and how the rollover occurs." In describing this test procedure, Larson stated: "A wire rope tow system accelerates the dolly to the desired speed. As the Roller Coaster Dolly approaches the release point, the rail descends into a trough so that the top surface

FIGURE 1.4 A soil-tripped rollover test from Cooperrider [25].

of the dolly is level with the soil test surface. At the release point, the dolly is very quickly decelerated over a short distance by compressing aluminum honeycomb stacks against concrete blocks affixed in the ground. The vehicle then slides off the dolly and onto the test surface." Larson observed that this test procedure simulated the trip and roll phases of a rollover collision.

FIGURE 1.5 Roller coaster dolly.

In his paper, Larson presented data from two tests that utilized the Roller Coaster Dolly, one with a 1991 Jeep Wrangler and another with a 1986 Chevrolet S-10 Blazer. A scale diagram of the initial and final vehicle positions and physical evidence from the Jeep Wrangler test is included in Figure 1.6. Each vehicle was released with a slip angle of approximately 52° across a dirt surface with a 15.5° downslope. The vehicles deposited tire and rim furrows in the dirt before beginning to roll over. Larson reported that the Jeep Wrangler was traveling 40.7 mph (65.5 kph) when it was released from the dolly, that it traveled 70.5 ft (21.5 m) from release to rest, that it furrowed for approximately 20.7 ft (6.3 m) before entering the roll phase, and that it completed one roll before coming to rest. The roll distance was approximately 49.8 ft (15.2 m). Larson did not report a speed for the vehicle at the end of the furrowing. Larson reported that the Chevrolet S-10 Blazer was traveling 41.1 mph (66.1 kph) when it was released from the dolly, that it traveled 74 ft (22.6 m) from release to rest, that it furrowed for approximately 18.1 ft (5.5 m) before entering the roll phase, and that it completed two rolls before coming to rest. The roll distance was approximately 55.9 ft (17.1 m). Again, Larson did not report a speed for the vehicle at the end of furrowing.

In 2002, Hughes [1] reported a series of 8 curb-tripped rollover tests conducted with 2001 Nissan Pathfinders. These tests used a modified dolly fixture that did not introduce an initial roll angle to the test vehicles. Each vehicle was launched laterally from the dolly with a speed of around 31 mph (50 kph), and then the leading-side tires impacted a curb, which was approximately 6 in. (15 cm) high. Rollover occurred in all 8 tests. The roll distances ranged from 54 to 72 ft (16.5 to 21.9 m), and the number of rolls varied between 1 and 2½ times. Peak accelerations during the curb impacts ranged from 14.5 to 20.5 g. Hughes noted that "small changes in boundary conditions, such as angle at ground contact, resulted in major differences during the dynamic event." Figure 1.7 depicts a series of sample vehicle positions from two of these tests. These positions demonstrate what is true of every rollover test procedure where the vehicle is allowed to roll freely after initiation - that regardless of how it is initiated, the roll phase of the test is not repeatable. Small differences in the initial conditions lead to large differences in the rollover motion.

FIGURE 1.6 Rollover test from the rollover coaster dolly.

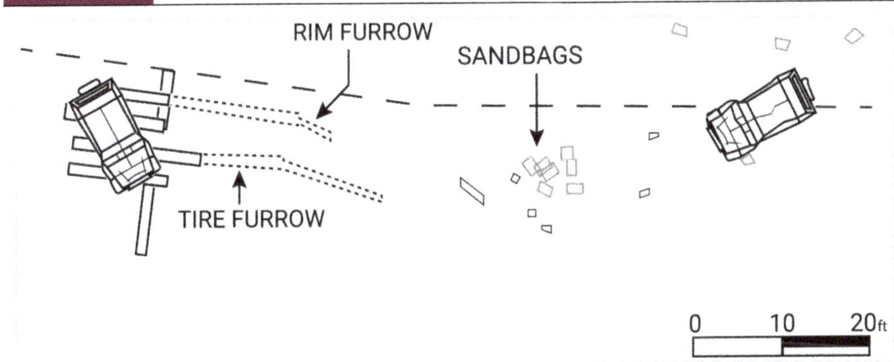

FIGURE 1.7 Sample roll motion from the Hughes Nissan Pathfinder curb-tripped tests.

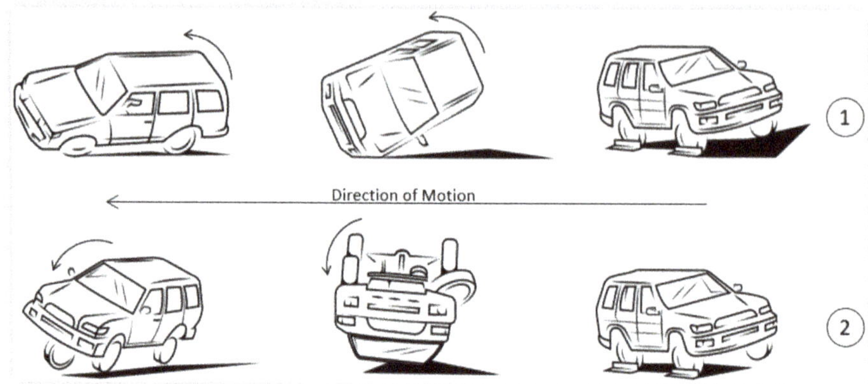

Rollover Tests Using Automated Steering Control

In 1994, Nalecz and Lu [27] reported that they ran over 150 full-scale vehicle rollover initiation tests using 8 different vehicles (2 passenger cars, 2 pickup trucks, 2 SUVs, and 2 vans) equipped with automated steering control. The vehicles were tested under 2 different loading conditions. Nalecz and Lu reported that "during testing all vehicles were guided by automatic vehicle controller which had been especially designed for this project. The controller accelerated [the] vehicle from rest to a desired speed along [a] straightaway, applied the steering input and then coast the vehicle along the curve until brakes were applied to generate vehicle spinout. The brake were applied simultaneously to both rear wheels and the inner front wheel in order to quickly generate the highest possible yaw rate." Some of these tests involved both a loss of control and a trip. The vehicles were tested with outriggers that arrested the roll motion of the vehicle after they started to tip up.

Nalecz and Lu processed sensor data from 77 of these tests. These 77 tests included 42 unloaded vehicles and 35 loaded vehicles. In 56 of the tests, the vehicles did not tip up, while in 21 tests they did. No tip-ups occurred with the passenger vehicles. Based on their testing results, Nalecz and Lu observed that "vehicles tested are easier to roll over when they are tripped by a curb than when skidded into soil." Also, "these results… indicate that vehicle lateral velocity contributes more to vehicle rollover than forward velocity." Beyond these basic observations, Nalecz and Lu did not present their results in a manner useful to accident reconstructionists and their tests are not treated further in this book.

In 2000, Larson [26] also reported a second test procedure that used a cable guide and tow system to accelerate the vehicle up to the test speed. The vehicle was then released onto an asphalt surface where an automated steering-controller-applied steer input to the test vehicle. In his paper, Larson reported test data from one test that resulted in a rollover. This test utilized a Toyota 4WD pickup with the following steering inputs: 160° to the right at a steering rate of 800°/s, 180° to the left at a steering rate of 755°/s, and 350° to the right at a steering rate of 964°/s. Larson reported that the lateral acceleration of the vehicle reached 0.7 g during the first steer, 0.8 g during the second, and 1.07 g during the third steer. The test vehicle had been modified with aftermarket wheels and tires that were larger than the OEM wheels and tires, and ballast had been added to the vehicle. In this test, the vehicle was traveling 78 mph (126 kph) when it was released from the guide and tow system. Larson reported that the vehicle traveled approximately

FIGURE 1.8 Rollover test by Larson.

355 ft (108 m) before it became airborne. It then rolled for 260 ft (80 m), completing seven rolls before coming to rest. Figure 1.8 depicts the Toyota pickup rollover. Arndt [28] analyzed Larson's data and reported that the speed of the vehicle was 57 mph (92 kph) at the beginning of the roll phase.

In two articles published in 2007 in *Collision*, Wilson [29, 30] described two steering-induced rollover crash tests. The first test utilized a 1991 Ford Explorer XLT. The rollover of this vehicle was preceded by remotely controlled steering inputs of 240° to the left and then 355° to the right. Both steering inputs were conducted at a rate of 600°/s. The Explorer deposited 143 ft (43.6 m) of tire marks and then began rolling over, completing four rolls in 164 ft (50 m). The speed of the vehicle at the beginning of the rollover was approximately 45 mph (72.6 kph).

The second test utilized a 1997 Toyota 4Runner SR5. The rollover of this vehicle was preceded by steering inputs of 360° to the left and then 360° to the right. Both steering inputs were conducted at a rate between 900 and 1000°/s. The 4Runner "experienced a driver's side-leading rollover. It rolled ¾ of a roll before the rollover motion stalled at the end of Roll 1. As the 4Runner was rolling onto its passenger's side tires, the vehicle had yawed such that it was traveling backwards. It traveled backwards on its passenger's side tires for approximately 70 ft (21.3 m). At this point, the 4Runner started to reverse its yaw and roll motion. The 4Runner yawed counterclockwise (CCW) from an approximate 180° orientation relative to its initial direction of travel to a 90° CW orientation. While the 4Runner was reversing its yaw direction back to 90°, it reversed its roll motion from 7/8 of a roll (traveling on its passenger's side tires) to ¾ of a roll (falling back onto its passenger's side). Once on its passenger's side, with the vehicle's orientation approximately 90° relative to its direction of travel, it started a second driver's side-leading rollover sequence. This sequence lasted for one more complete roll, starting at ¾ of a roll and ending at 1¾ rolls."

TABLE 1.2 Summary of the tests in Asay's 2009 article

Study	Test device	Trip mechanism	Roll surface type	Vehicle type	Year	Make	Model	Release speed		Speed et beginning of roll		Roll distance		Number of rolls
								mph	kph	mph	kph	ft	m	
Asay [33]	Steering Controller	-	-	Passenger Car	1996	BuicK	Skylark	75	120.7	-	-	-	-	0
	Steering Controller	Soil	Soil and Asphalt	Car	1996	Buick	Skylark	75	120.7	41.6	66.9	122.4	37.3	6
	Steering Controller	Soil	Soil	Passenger Car	1984	AMC	Eagle	75	120.7	34.6	55.7	82.2	25.1	3
	Steering Controller	-	-	Passenger Car	1987	Ford	Taurus	75	120.7	-	-	-	-	0
	Steering Controller	Soil	Soil	Passenger Car	1987	Ford	Taurus	75	120.7	27.1	43.6	62.5	19.1	1
	Steering Controller	Soil/ Asphalt	Asphalt	Passenger Car	1991	Ford	Escort	75.7	121.83	51.6	83.0	173.6	52.9	8

In 2009, Asay [31] reported six full-scale tests of passenger cars that utilized automated steering control (summarized in Table 1.2).[8] Four of these tests resulted in rollover and two did not. Five of these six tests were conducted on a remote, rural highway in the west desert of Utah. This roadway was relatively level and consisted of a lane in each direction with a chip-seal-coated surface and graded shoulders along both sides. Asay reported that the friction coefficient for this highway was 0.795 (measured using a vehicle equipped with a Vericom performance measurement device). The off-road terrain surrounding this highway was open desert with sagebrush and sparse grass. The vehicles in these five tests were shifted into neutral (without the engine running) and then were towed up to speed using a modified 1986 International conventional ten-wheeled tractor (the Massive Moving Barrier (MMB) used by Asay in previous testing; see Figure 1.9). They were then released and the automated steering controller was used to input a series of steering inputs designed to lead to rollover. This occurred in three of the five tests. The tests in this series that resulted in rollover were steering induced, thus they provide data relevant to all three phases of high-speed, single-vehicle rollover accidents - the loss-of-control, the trip, and the roll.

The sixth test was conducted on a surface initially consisting of a single lane of asphalt with a dirt right shoulder. This lane with a shoulder was followed by a flat asphalt surface. The test vehicle in this sixth test was towed up to speed and then released. An automated steering controller was used to steer the vehicle. The vehicle was steered to the right, causing it to enter the dirt shoulder. It was then steered aggressively back

FIGURE 1.9 The Massive Moving Barrier (MMB) used to tow the test vehicles up to speed.

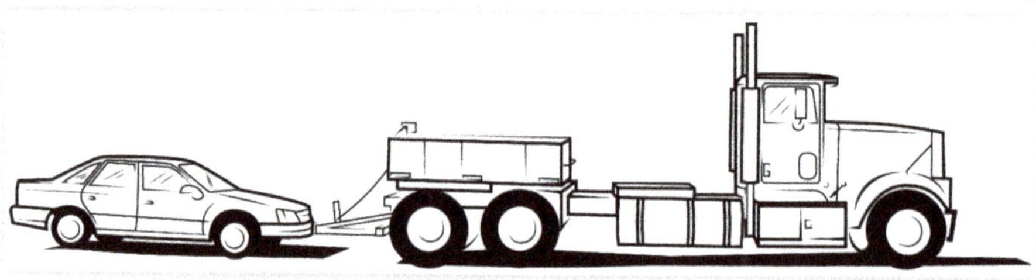

[8] The speeds reported in Table 1-2 for the beginning of the roll phase are from Arndt's 2011 study in which he made corrections to the speeds originally reported by Asay.

FIGURE 1.10 Vehicle motion from one of Asay's 2009 tests.

to the left, causing it to yaw, trip, and roll. The vehicle rolled across the asphalt surface. Figure 1.10 shows sample frames of video from this sixth test.

In 2010, Asay [32] reported another series of steering-induced rollover crash tests that were again conducted on the same remote, rural highway in the west desert of Utah, this time with six full-size SUVs of differing makes and models. These tests are summarized in Table 1.3. Asay's article reported 8 total tests, but two of them did not result in rollover. These tests again utilized automated steering control. Each vehicle was shifted into neutral and towed to highway speeds, released, and then steered with a sharp left steering input of approximately ¼ turn. One second later, the steering controller input steer back to the passenger's side until the limit was reached. Speed and position data was collected for each test vehicle using a VBOX III GPS sensor that was attached to the roof at the centerline and that collected data at a rate of 100 Hz. A noncontact radar speed sensor was also attached to the rear bumper, and a rotation rate sensor was attached near the center of gravity to measure the roll velocity. Additional instrumentation is discussed in Asay's article. Asay documented the physical evidence from each of the tests that he reported in 2009 and 2010. He noted that this physical evidence included tire marks, tire furrows, gouges, wheel landings, glass spills, scrapes, and debris from the vehicle. For each test, this physical evidence was located using a Leica total station and an evidence diagram was produced in a CAD package using the survey data. Like Asay's previous set of tests, these tests were steering induced and utilized steering inputs achievable by a human driver, and thus, they provide data relevant to all three phases of high-speed, single-vehicle rollover accidents - the loss-of-control, the trip, and the roll. Figure 1.11 contains frames of video from one of Asay's 2010 tests.

In 2011, Stevens reported analysis of five full-scale tests in which a vehicle was driven or towed to highway speed then steered with sufficient severity to induce a rollover. The double-step steer inputs were generated using a programmable remote control radio. Tests were run with four SUVs and one minivan. In all five tests, the steering maneuvers began on the paved surface. Some of the vehicles tripped on the pavement, and others left the paved surface and tripped on a soil surface. Stevens documented the physical evidence from each of these tests with a survey and photographs. The physical evidence on each vehicle (scratch patterns and crush damage) was also documented. These tests are summarized in Table 1.4.

In 2015, Asay, Carter, Funk, and Stephens reported another full-scale, steering-induced rollover crash test involving a 1996 GMC Jimmy 4x4 SUV. This test was conducted at the Wendover Airport in western Utah. As in previous tests reported by Asay, the test vehicle was towed up to speed using the MMB test vehicle, which is a

TABLE 1.3 Summary of the tests in Asay's 2010 article

Study	Test device	Trip mechanism	Roll surface type	Vehicle type	Year	Make	Model	Release speed		Speed at beginning of trip		Trip distance		Speed at beginning of roll		Roll distance		Number of rolls
								mph	kph	mph	kph	ft	m	mph	kph	ft	m	
Asay [33]		Soil	Soil and Asphalt	SUV	1996	Oldsmobile	Bravada	75.7	121.8	-	-	-	-	57.2	92.1	266.6	81.3	10 1/2
	Steering Controller	Soil	Soil	SUV	1991	Isuzu	Rodeo	73.5	118.3	-	-	-	-	44.4	71.5	177	53.9	7
	Steering Controller	-	-	SUV	1994	Nissan	Pathfinder	63.9	102.8	-	-	-	-	-	-	-	-	0
	Steering Controller	Soil	Soil and Asphalt	SUV	1994	Nissan	Pathfinder	69.9	112.5	-	-	-	-	36.7	59.1	114.1	34.8	3 1/2
	Steering Controller	-	-	SUV	2002	Ford	Explorer	60.1	96.7	-	-	-	-	-	-	-	-	0
	Steering Controller	Soil	Soil	SUV	2002	Ford	Explorer	70.0	112.7	-	-	-	-	32.4	52.1	78.6	24.0	2
	Steering Controller	Soil	Soil	SUV	1998	Ford	Expedition	70.0	112.7	-	-	-	-	28.7	46.2	52.5	16.0	1 1/2
	Steering Controller	Soil	Soil and Asphalt	SUV	1991	Mitsubishi	Montero	74.6	120.1	-	-	-	-	61.0	98.2	242.6	73.9	8 1/2

FIGURE 1.11 Sample vehicle motion from Asay's 2010 Test #2.

TABLE 1.4 Summary of Stevens' rollover tests reported in 2011

Study	Test device	Trip mechanism	Roll surface type	Vehicle type	Year	Make	Model	Release speed		Speed at beginning of roll		Roll distance		Number of rolls
								mph	kph	mph	kph	ft	m	
Stevens [34]	Steering Controller	Steering	Concrete	Minivan	1989	Ford	Aerostar	-	-	33.8	54.4	104.5	31.9	2 1/2
	Steering Controller	Steering	Concrete and Soil	SUV	2001	Chevrolet	Blazer	-	-	57.9	93.2	264.7	80.7	9
	Steering Controller	Steering	Concrete and Soil	SUV	2002	Ford	Explorer Sport 4x2	-	-	41.0	66.0	146.7	44.7	5
	Steering Controller	Soil	Soil	SUV	1997	Ford	Explorer Sport 4x4	-	-	30.2	48.6	92.4	28.2	3 1/2
	Steering Controller	Soil	Soil	SUV	1995	Nissan	Pathfinder	-	-	42.3	68.1	105.2	32.1	3

structurally reinforced 1986 International ten-wheeled tractor weighing approximately 43,000 lb. An inertia-based internal measurement unit (IMU) that utilized GPS and internal integration was mounted inside the test vehicle along with additional high-speed accelerometers for comparison purposes. Additionally, the test was documented with 4 stationary, 2 panning, and 5 moving cameras (including video from an overhead UAV). The vehicle transmission was shifted into neutral during the test and the engine was not running. Prior to the test, the windows, mirror, and turn signal on the driver's side of the vehicle were painted yellow for later posttest vehicle roll identification. The passenger's side was painted orange and the rear glass was painted green for the same purpose. The GMC Jimmy was towed to a speed of 53.6 mph (86.3 kph) and then released. A steering controller then steered the vehicle through a series of steering inputs that led to the vehicle rolling over. The vehicle rolled 2 complete revolutions. This rollover was soil-tripped on a relatively flat, but naturally occurring, surface.

In another 2015 study, Larson and his colleagues [35] reported two additional soil-tripped rollover tests for which the loss of control was generated with an automated steering controller. These tests were run at Exponent's Test and Engineering Center

FIGURE 1.12 Aerial view of Exponent's Test and Engineering Center.

(TEC), located in Phoenix, Arizona. The tests utilized a crew-cab pickup and a midsized sedan and involved the vehicles being steered off the paved roadway onto a soil surface. Figure 1.12 depicts an aerial view of the test area. This area included an asphalt tow lane that led from a crash rail to an asphalt skid pad. The soil test surface has been prepared to the right of the tow lane. The initial steer of each test vehicle was to the right and then, once the vehicle was on the soil surface, a countersteer to the left generated a counter-clockwise yaw that resulted in furrowing and then rollover. The vehicles returned to the paved roadway during portions of the roll phase. During the tests, the vehicle engines were running at idle and the transmissions were in neutral. The electronic stability control on each vehicle was deactivated prior to the tests.

The initial test speed for the pickup was 72 mph. After the steering inputs and the furrowing of the vehicle tires into the soil, the speed of the pickup had been reduced to 44 mph at the beginning of the rollover phase. The trip phase occurred over a distance of 58 ft, during which the lateral accelerations reached 2.4 g and the longitudinal accelerations reached 1.7 g. The pickup completed four passenger-side-leading rolls over a distance of 144 ft, coming to rest on its wheels.

The initial test speed for the sedan was 65 mph. After the steering inputs and the furrowing of the vehicle tires into the soil, the speed of the sedan had been reduced to 34 mph at the beginning of the rollover phase. The trip phase occurred over a distance of 56 ft, during which the ground plane accelerations reached 3.5 g. The pickup completed 3½ passenger-side-leading rolls over a distance of 95 ft, coming to rest on its roof.

Controlled Rollover Impact System

In 2001, Cooper [36] reported a new test procedure, called the Controlled Rollover Impact System (CRIS) in which the conditions of the first roof-to-ground impact could be precisely controlled. This was accomplished with a drop fixture attached to a moving semitrailer that could rotate and release a vehicle onto the ground. Figure 1.13 depicts the test fixture with an SUV attached to it. This fixture allows control of the yaw, pitch, and roll angles at the first roof-to-ground impact, along with the translational velocity, the vertical velocity, and the roll velocity. Once the vehicle is released from the fixture and the roof strikes the ground, the vehicle is allowed to continue uninterrupted to rest.

FIGURE 1.13 CRIS test fixture.

Carter et al. [37] compared the CRIS test fixture and procedure to the traditional dolly rollover test. To accomplish this, a dolly rollover test was run with a Ford Econoline 350 van. The van was released from the dolly at a speed of 55.3 mph (88.5 kph), with its passenger side leading. They then analyzed this test to determine the initial conditions for the first roof-to-ground impact that occurred during the test. They then used these conditions as the initial conditions for the CRIS test, with another identical van. These conditions were as follows: the initial pitch and yaw angles were 0°, the vehicle was translating at a speed of 51 mph (82.1 kph), the vehicle had an initial roll rate of 149°/s, the drop height was set to 5 in., and the target roll angle at impact with the ground was 133°. Carter, Habberstad, and Croteau reported that "the observed vehicle dynamics demonstrated a dramatic difference between the two test devices given similar initial conditions of the first roof-to-ground impact." In addition to this comparison, the authors quantified the differences in accelerations between the vehicle's center of gravity and the roof rails during the roof-to-ground impacts. They concluded that "during roof-to-ground engagements, the environment near the roof rails is much more severe, in terms of acceleration, than that observed at the vehicle's center of gravity."

Moffatt [38] used the CRIS test fixture to test three vehicles with roll cages and three with production roofs in matched-pair roof-to-ground impacts. The vehicles contained seat-belted dummies, and Moffatt was exploring the degree to which roof strength influenced the loads on the occupants. He found that "the seat belted dummies measured nearly identical head impacts and neck loads with or without the roll cage, despite significant roof crush in the production roof vehicles. Roof crush had no measureable influence on the severity of the head accelerations and neck loads." Moffatt's testing benefitted from the CRIS test fixture because of the control it provides over the roof-to-ground impact conditions.

Real-World Rollovers

In addition to staged rollover tests, researchers can also learn much from real-world rollover crashes that have been captured on video. In a 2007 article, Rose and Beauchamp [39] analyzed the dynamics of 12 high-speed, real-world rollover crashes that were captured on video. Their study focused primarily on the vehicle roll velocities and the characteristics of typical roll velocity curves. For the twelve rollover accidents analyzed, the analysis yielded roll rates at the beginning of the roll phase between 150 and 540°/s. Peak roll rates varied between 300 and 675°/s, and average roll rates varied between 144 and 395°/s. The number of rolls and the initial, peak, and average roll rates for the rollovers analyzed in this study are summarized below in Table 1.5. Figure 1.14 shows frames from the video of Case #3, and Figure 1.15 shows frames from the video of Case #4.

In 2008, Anderson [40] presented analysis of a real-world rollover crash involving a 2003 Ford Explorer Sport that began rolling on the roadway and continued rolling off the roadway. The crash was captured by the video camera in a law enforcement vehicle. Anderson documented the vehicle and the crash site and then analyzed the video, concluding that the vehicle was initially traveling 79 mph (127.4 kph) and that, at the beginning of the roll phase, it was traveling 54 mph (87.1 kph). The vehicle completed six rolls, the average roll velocity for the roll phase exceeded 300°/s, with a peak value of approximately 540°/s. Anderson used the video to determine the deceleration rate of the vehicle during the roll and found that it varied from approximately 0.58 g early in the roll to 0.2 g near the end. The average deceleration without adjusting for grade was 0.39 g. After adjusting for grade, it was 0.43 g.

TABLE 1.5 Summary of the real-world rollovers analyzed by Rose and Beauchamp

Study	Test device	Trip mechanism	Vehicle type	Number of rolls	Initial roll velocity (deg/s)	Average roll velocity (deg/s)	Peak roll velocity (deg/s)	Naturally occurring?
Rose and Beauchamp [39]	Real-World Rollover	Off-Road Furrow	Pickup	2	245	247	386	Yes
	Real-World Rollover	On-Road	SUV	1 3/4	270	338	415	Yes
	Real-World Rollover	Off-Road Furrow	SUV	3 3/4	208	302	450	Yes
	Real-World Rollover	Impact	Rally Car	4	540	272	573	Yes
	Real-World Rollover	Off-Road Furrow	Road Race Car	2 1/4	338	173	338	Yes
	Real-World Rollover	Impact	Rally Car	3	169	261	476	Yes
	Real-World Rollover	Ramping	Rally Car	5	169	310	491	Yes
	Real-World Rollover	Drop Off Road	Rally Car	4	225	174	386	Yes
	Real-World Rollover	Impact	Rally Car	1 1/2	270	190	300	Yes
	Real-World Rollover	Impact	Rally Car	1 1/2	225	144	300	Yes
	Real-World Rollover	Dirt Road Furrow	Rally Car	7	150	395	600	Yes
	Real-World Rollover	Off-Road Furrow	Rally Car	4	-	-	-	Yes

FIGURE 1.14 Case #3 from Rose and Beauchamp [39].

FIGURE 1.15 Case #4 from Rose and Beauchamp [39].

References

1. Hughes, R., Lewis, L., Hare, B., Ishikawa, Y. et al., "A Dynamic Test Procedure for Evaluation of Tripped Rollover Crashes," SAE Technical Paper 2002-01-0693, 2002, doi:10.4271/2002-01-0693.
2. Rose, N., Carter, N., and Beauchamp, G., "Post-Impact Dynamics for Vehicles with a High Yaw Velocity," SAE Technical Paper 2016-01-1470, 2016, doi:10.4271/2016-01-1470.
3. MacInnis, D., Cliff, W., and Ising, K., "A Comparison of Moment of Inertia Estimation Techniques for Vehicle Dynamics Simulation," SAE Technical Paper 970951, 1997, doi:10.4271/970951.
4. Garrott, W.R., "Inertial Parameters of Selected Four Wheel Drive Utility Vehicles," Project VRTC-88-087, Report #1, National Highway Traffic Safety Administration (NHTSA), Washington, DC, 1988.
5. Garrott, W., Monk, M., and Chrstos, J., "Vehicle Inertial Parameters-Measured Values and Approximations," SAE Technical Paper 881767, 1988, doi:10.4271/881767.
6. Garrott, W., "Measured Vehicle Inertial Parameters -NHTSA's Data Through September 1992," SAE Technical Paper 930897, 1993, doi:10.4271/930897.
7. Heydinger, G., Bixel, R., Garrott, W., Pyne, M. et al., "Measured Vehicle Inertial Parameters-NHTSA's Data through November 1998," SAE Technical Paper 1999-01-1336, 1999, doi:10.4271/1999-01-1336.
8. Allen, R., Klyde, D., Rosenthal, T., and Smith, D., "Estimation of Passenger Vehicle Inertial Properties and Their Effect on Stability and Handling," SAE Technical Paper 2003-01-0966, 2003, doi:10.4271/2003-01-0966.
9. Garrott, W., "The Variation of Static Rollover Metrics with Vehicle Loading and between Similar Vehicles," SAE Technical Paper 920583, 1992, doi:10.4271/920583.

10. Shapiro, S., Dickerson, C., Arndt, S., Arndt, M. et al., "Error Analysis of Center-of-Gravity Measurement Techniques," SAE Technical Paper 950027, 1995, doi:10.4271/950027.
11. Achhammer, G., Frimberger, M., Glaser, T., Wolf, F. et al., "Vehicle Tests and Simulation Tools for Rollover Sensing," SAE Technical Paper 1999-01-0438, 1999, doi:10.4271/1999-01-0438.
12. Wilson, R. and Gannon, R., "Rollover Testing," SAE Technical Paper 720495, 1972, doi:10.4271/720495.
13. SAE International Surface Vehicle Information Report, "Rollover Test Methods," SAE Standard J2926, Aug. 2011.
14. SAE Recommended Practice, "Dolly Rollover Recommended Test Procedure," SAE Standard J2114, Feb. 2011.
15. Orlowski, K.F., Bundorf, R.T., and Moffatt, E.A., "Rollover Crash Tests-The Influence of Roof Strength on Injury Mechanics," SAE Technical Paper 851734, 1985, doi:10.4271/851734.
16. Orlowski, K., Moffatt, E., Bundorf, R., and Holcomb, M., "Reconstruction of Rollover Collisions," SAE Technical Paper 890857, 1989, doi:10.4271/890857.
17. Bahling, G.S., Bundorf, R.T., Kaspzyk, G.S., Moffatt, E.A. et al., "Rollover and Drop Tests-The Influence of Roof Strength on Injury Mechanics Using Belted Dummies," SAE Technical Paper 902314, 1990, doi:10.4271/902314.
18. Luepke, P., Carhart, M., Croteau, J., Morrison, R. et al., "An Evaluation of Laminated Side Window Glass Performance During Rollover," SAE Technical Paper 2007-01-0367, 2007, doi:10.4271/2007-01-0367.
19. Luepke, P., Carter, J., Henry, K., Germane, G. et al., "Rollover Crash Tests on Dirt: An Examination of Rollover Dynamics," *SAE Int. J. Passeng. Cars - Mech. Syst.* 1, no. 1 (2009): 18-30, doi:10.4271/2008-01-0156.
20. Rose, N., Neale, W., Fenton, S., Hessel, D. et al., "A Method to Quantify Vehicle Dynamics and Deformation for Vehicle Rollover Tests Using Camera-Matching Video Analysis," *SAE Int. J. Passeng. Cars - Mech. Syst.* 1, no. 1 (2009):301-317, doi:10.4271/2008-01-0350.
21. Yaek, J., Curry, B., and Goertz, A., "Review and Comparison of Published Rollover Test Results," SAE Technical Paper 2010-01-0057, 2010, doi:10.4271/2010-01-0057.
22. Croteau, J., Zolock, J., Larson, R., Bare, C. et al., "Dynamic Response of Vehicle Roof Structure and ATD Neck Loading during Dolly Rollover Tests," *SAE Int. J. Passeng. Cars - Mech. Syst.* 3, no. 1 (2010): 407-449, doi:10.4271/2010-01-0515.
23. Turriff, D., King, D., and Bertoch, J., "On the Directionality of Rollover Damage and Abrasions," SAE Technical Paper 2015-01-1421, 2015, doi:10.4271/2015-01-1421.
24. Cooperrider, N., Thomas, T., and Hammoud, S., "Testing and Analysis of Vehicle Rollover Behavior," SAE Technical Paper 900366, 1990, doi:10.4271/900366.
25. Cooperrider, N., Hammoud, S., and Colwell, J., "Characteristics of Soil-Tripped Rollovers," SAE Technical Paper 980022, 1998, doi:10.4271/980022.
26. Larson, R., Smith, J., Werner, S., and Fowler, G., "Vehicle Rollover Testing, Methodologies in Recreating Rollover Collisions," SAE Technical Paper 2000-01-1641, 2000, doi:10.4271/2000-01-1641.
27. Nalecz, A. and Lu, Z., "Methodology for Tripped Vehicle Rollover Testing and Analysis of Experimental Results," SAE Technical Paper 940225, 1994, doi:10.4271/940225.

28. Arndt, M., "Drag Factors from Rollover Crash Testing for Crash Reconstructions," *Proceedings of the ASME 2011 International Mechanical Engineering Congress & Exposition IMECE 2011*, Denver, CO, November 11-17, 2011, IMECE2011-65537.

29. Wilson, L.A., Gilbert, M., and Godrick, D., "Reconstruction and Analysis of Steering-Induced, On-Road, Untripped SUV Rollover Tests (Part 1)," *Collision—The International Compendium for Crash Research* 2, no. 1, Spring 2007.

30. Wilson, L.A., Gilbert, M., and Godrick, D., "Reconstruction and Analysis of Steering-Induced, On-Road, Untripped SUV Rollover Tests (Part 2)," *Collision—The International Compendium for Crash Research* 2, no. 2, Fall 2007.

31. Asay, A. and Woolley, R., "Rollover Testing on an Actual Highway," *SAE Int. J. Passeng. Cars - Mech. Syst.* 2, no. 1 (2009): 1237-1267, doi:10.4271/2009-01-1544.

32. Asay, A., Carter, J., Funk, J., and Stephens, G., "Rollover Testing of a Sport Utility Vehicle (SUV) with an Inertial Measurement Unit (IMU)," SAE Technical Paper 2015-01-1475, 2015, doi:10.4271/2015-01-1475.

33. Asay, A. and Woolley, R., "Rollover Testing of Sport Utility Vehicles (SUVs) on an Actual Highway," SAE Technical Paper 2010-01-0521, 2010, doi:10.4271/2010-01-0521.

34. Stevens, D., Arndt, S., Wayne, L., Arndt, M. et al., "Rollover Crash Test Results: Steer-Induced Rollovers," SAE Technical Paper 2011-01-1114, 2011, doi:10.4271/2011-01-1114.

35. Larson, R., Croteau, J., Bare, C., Zolock, J. et al., "Steering Maneuver with Furrow-Tripped Rollovers of a Pickup and Passenger Car," SAE Technical Paper 2015-01-1477, 2015, doi:10.4271/2015-01-1477.

36. Cooper, E., Moffatt, E., Curzon, A., Smyth, B. et al., "Repeatable Dynamic Rollover Test Procedure with Controlled Roof Impact," SAE Technical Paper 2001-01-0476, 2001, doi:10.4271/2001-01-0476.

37. Carter, J., Habberstad, J., and Croteau, J., "A Comparison of the Controlled Rollover Impact System (CRIS) with the J2114 Rollover Dolly," SAE Technical Paper 2002-01-0694, 2002, doi:10.4271/2002-01-0694.

38. Moffatt, E., Cooper, E., Croteau, J., Orlowski, K. et al., "Matched-Pair Rollover Impacts of Rollcaged and Production Roof Cars Using the Controlled Rollover Impact System (CRIS)," SAE Technical Paper 2003-01-0172, 2003, doi:10.4271/2003-01-0172.

39. Rose, N., Beauchamp, G., and Fenton, S., "Factors Influencing Roof-to-Ground Impact Severity: Video Analysis and Analytical Modeling," SAE Technical Paper 2007-01-0726, 2007, https://doi.org/10.4271/2007-01-0726.

40. Anderson, J., Gee, R., Germane, G., Henry, K. et al., "Analysis of a Real-World High-Speed Rollover Crash from a Video Record and Physical Evidence," SAE Technical Paper 2008-01-1486, 2008, doi:10.4271/2008-01-1486.

CHAPTER 2

Physical Evidence from Rollover Crashes

This chapter explores the types of physical evidence that can be deposited on the ground and damage that can be sustained by a vehicle during a rollover crash. The first section focuses on evidence deposited on the ground (scene evidence), and the second focuses on the damage sustained by the vehicle (vehicle evidence). The third section addresses applications of photogrammetry, a set of techniques commonly used for documenting and diagramming physical evidence from scenes and vehicles. While a bit melodramatic, it is true what Cofone [1] has written about physical evidence: "When you arrive at the scene of any traffic crash, the physical evidence, injury patterns and damage profiles of the vehicles scream of a story. The cries are so loud they are almost deafening. Where we sometimes go awry is in not knowing how to listen to the story." This chapter is aimed at developing our ability to listen effectively to the story of what actually occurred when we examine the physical evidence from a rollover.

Scene Evidence

Scene evidence typically needs to be documented and preserved for later use in the reconstruction. This process involves three phases: identification, documentation, and quantification. Identification involves recognizing the presence of a piece of evidence, either at the accident site or in photographs, and then determining if the evidence is actually from the crash under investigation. In the early stages of evidence preservation, it may not be apparent whether or not a piece of evidence is from the subject crash. In such cases, it is typically better to capture the evidence and to rule it in or out during later analysis. The identification process also typically includes marking the evidence to distinguish it from the natural surroundings by utilizing chalk, paint, flags, or cones.

During the documentation phase, the investigator captures the evidence within the context of its surroundings using a camera to capture photographs and video. The use of unmanned aerial vehicles (UAVs) or elevated cameras can provide perspectives that are useful in later analysis. Nearby bridges or hillsides can also be used to gain these elevated vantage points for taking photographs or video. During the quantification phase, the evidence is measured and located within its surroundings using tape measures, a measurement wheel, a total station, a laser scanner, or through photogrammetry using photographs or video captured with a UAV.

As an example, consider Figure 2.1, which is a scene evidence diagram produced based on posttest documentation of the test area following the Isuzu Rodeo rollover test reported by Asay in 2010. This diagram is oriented such that the vehicle would have moved from left to right across the diagram. The diagram of Figure 2.1 depicts certain features of the roadway along with the specific locations of evidence deposited by the vehicle during the test. The physical evidence depicted in Figure 2.1 includes tire marks, furrows, disturbed earth, glass, and pieces that broke off of the vehicle during the rollover.

Figure 2.1 is the type of scale diagram that could be used to reconstruct the motion of the vehicle. This reconstruction process could be carried out in a computer modeling program (such as AutoCAD) where a computer model of the vehicle could be placed in the locations and orientations necessary to deposit the documented physical evidence. Alternatively, this diagram could be printed out to the scale of a physical model of the vehicle, and the reconstructionist could use the physical model to evaluate the location and orientation of the vehicle needed to deposit the documented physical evidence.

Now, consider another one of the eight tests that Asay reported in 2010 (Test #6). This test, conducted on August 19, 2009, utilized a 2002 Ford Explorer. The vehicle was released at a speed of 70 mph (113 kph) and was steered first to the left (driver's side) approximately ¼ of a turn of the steering wheel and then aggressively back to the right (passenger's side) until the rightward steering stop was reached. The vehicle exited the left side of the roadway in a clockwise yaw with its driver's side leading. After furrowing into the dirt and gravel and yawing past a 90° slip angle, the Ford began rolling over and completed 2 rolls in a distance of approximately 79 ft (24.1 m).

Asay noted that the physical evidence from this and other tests in his series included "tire marks, tire furrows, gouges, wheel landings, glass spills, scrapes, debris pieces, [and the] final rest position of the test vehicle...once the physical evidence was...marked, it was...measured by means of a Leica survey instrument (total station)." Asay provided the survey data that was used to create the evidence diagram included in Figure 2.2. This diagram shows tire marks, furrows, ground impact marks, and debris deposited by the vehicle during the loss-of-control, trip, and roll phases of this test. In this diagram, the roadway geometry and evidence have been overlaid on an aerial photograph.

FIGURE 2.1 Physical evidence diagram.

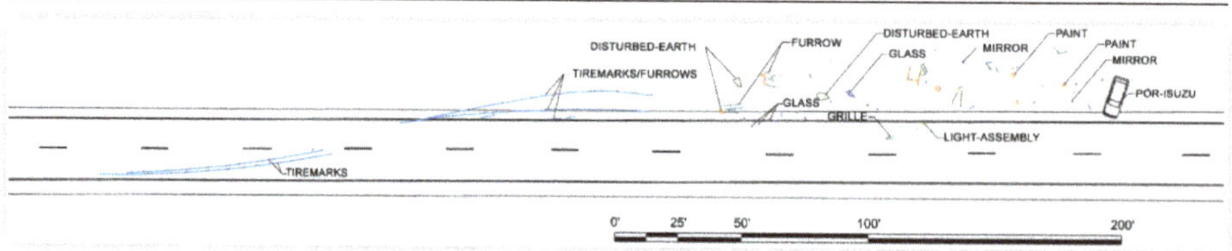

FIGURE 2.2 Evidence diagram for Test #6 from Asay [2].

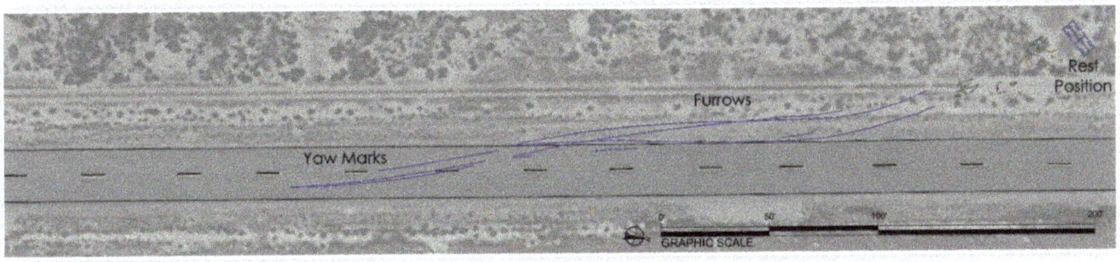

Figure 2.3 again depicts the scene evidence from this test, but this diagram includes vehicle positions that show the motion implied by the yaw marks and furrows. Ultimately, the purpose of a diagram like that of Figure 2.2 is to determine and present the motion of the vehicle based upon the previously documented physical evidence. The documented physical evidence provides the travel path of the vehicle and the specific characteristics of the motion along that path. In this case, the vehicle deposited tire marks on the roadway while it was yawing in a counterclockwise fashion following the leftward steering input and then in a clockwise fashion following the subsequent severe rightward steering input. The vehicle then entered the dirt shoulder while yawing in a clockwise direction with its driver's side leading and deposited furrows in the dirt. The distinct furrows in the dirt off the left side of the roadway are identified with orange flags in the photographs of Figures 2.4 through 2.6. In Figure 2.5, the point at which the trailing (passenger's side) tires lift off the ground and stop depositing furrows is visible. This represents the vehicle entering the trip phase. The rest position of the vehicle is also visible in these photographs.

The photographs of Figures 2.7 through 2.10 depict ground impact marks and debris from the roll phase of this test. These ground impact marks and debris have been identified with white spray paint in these photographs. These pieces of evidence, which are included in the diagram of Figure 2.2, would assist the reconstructionist in determining and describing the path of the vehicle from the end of the trip phase to the rest position. Often though, for real-world crashes, on-scene investigators do not document the locations of ground impact marks and debris in detail. It may be possible to add this detail later through photogrammetric analysis of scene photographs. This will sometimes be necessary because impact marks often deteriorate before later investigators are able to document them, and vehicle debris will often be moved or removed during cleanup of the scene. As is discussed below, glass deposits are often still in place when later investigators come on scene and, thus, can be valuable evidence of the path of the vehicle through the roll phase and of the roll position of the vehicle at discrete points along the roll path.

The site of this test was visited on December 9, 2015, six years and several months after the test was run. Figure 2.11 depicts the site at the time of this inspection. As this photograph shows, the asphalt had been removed and a dirt road remained. The grading of the shoulders appeared similar to what it was at the time of the test, but it was apparent

FIGURE 2.3 Vehicle motion for Test #6 as defined by tire marks and furrows.

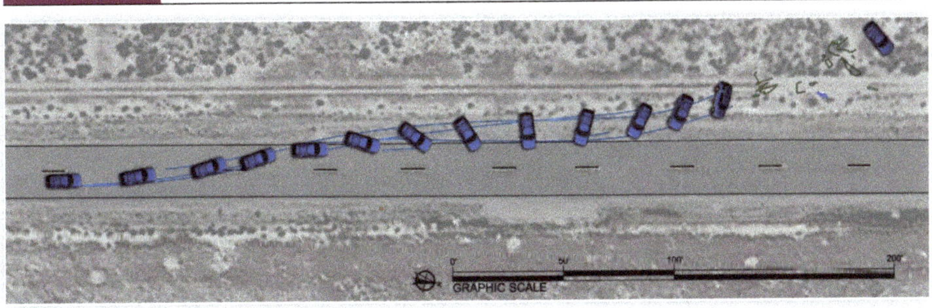

FIGURE 2.4 Photograph of tire furrows from a Ford Explorer.

FIGURE 2.5 Photograph of tire furrows from a Ford Explorer.

FIGURE 2.6 Photograph of tire furrows from a Ford Explorer.

FIGURE 2.7 Photograph of ground impact marks and debris from a Ford Explorer.

FIGURE 2.8 Photograph of ground impact marks from a Ford Explorer.

FIGURE 2.9 Photograph of ground impact marks from a Ford Explorer.

that the shoulders had been regraded during the removal of the asphalt. Chunks of asphalt from the original highway surface were present and mixed in with the soil on the shoulders. These alterations to the test site were more obvious than many of the scene alterations the authors have encountered in practice. Collisions investigators often need to consider subtle changes to crash sites. For instance, altered roadway stripping and

FIGURE 2.10 Photograph of ground impact marks from a Ford Explorer.

FIGURE 2.11 Site of Test #6 from Asay [2], 6 years and several months after the test.

changes to signage are commonplace. Such changes will be relevant if a reconstructionist intends to use techniques of photogrammetry in the course of their scene mapping, since such changes can prevent, or at least make more difficult, the use of stripping and signage as landmarks for evidence analysis.

The alterations to the roadway and shoulder in this particular instance make this an interesting case study in the longevity of particular pieces of scene evidence. Clearly, the tire marks on the roadway and furrows on the shoulder from this test were no longer present at the time of this inspection. However, three areas where vehicle window glass had been deposited on the ground were identifiable (Figure 2.12), as was vegetation that had been pressed down by the vehicle. The locations of individual bushes that were visible in the original photographs from the test that were still visible during this inspection made the rest position of the vehicle identifiable.

The photograph of Figure 2.13 shows a close-up view of glass fragments located in the area designated with the Number 2 in Figure 2.12. This glass exhibited dark tinting and

FIGURE 2.12 Glass debris fields identified with orange flags, Site of Test #6 from Asay [2], 6 years and several months after the test.

FIGURE 2.13 Close-up of glass debris from Area 2, site of Test #6 from Asay [2], 6 years and several months after the test.

many of the fragments were painted red on one side. The glass had a thickness of approximately 0.185 in.. Prior to this test, the left-side windows of the test vehicle were painted green, the right-side windows orange, and the rear hatch glass red. Thus, the red paint on the glass fragments in Area 2 identifies this glass as being from the rear hatch glass of the test vehicle.

The photograph of Figure 2.14 depicts a glass sample the authors collected from the area identified with a Number 3 in Figure 2.12. This glass exhibited dark tinting, a thickness of approximately 0.150 in., and some of the fragments had green paint on them. This green paint is an indication that this glass came from a left-side window of the test vehicle. Based on the thickness, this glass would be distinguishable from the rear hatch glass independent of this paint.

The glass in the area designated with the Number 1 in Figure 2.12 was clear tempered glass. This glass did not fit with any of the windows on the Ford Explorer from this test, all of which were painted. This glass is likely from one of Asay's 2009 tests, which were run in the same area, and in which the vehicle windows were not painted [3].

Glass deposits from broken vehicle windows can assist a reconstructionist in determining where during the roll phase a particular window broke and the roll orientation of the vehicle when it reached that point on the ground. The reconstructionist should be aware, though, that glass (particularly laminated glass and glass with film backing) can be thrown a considerable distance during the roll. For tempered glass, it will typically be clear, based on the shape and size of the debris field, whether the glass was deposited at a particular location or thrown there. The location of glass debris can also be essential for determining where during the roll phase an occupant was ejected or from which seating position that occupant was ejected [4, 5]. In order for an occupant to be ejected out of a particular window of the vehicle, the glass of that window must first break (unless the window is rolled down). Thus, knowing where along the roll path a particular window broke out will help place limits on when during the roll phase an occupant could have been ejected out of that window opening.

Perl [6] discussed the use of glass debris in the reconstruction of the roll phase. He observed that tempered glass that breaks and separates from the vehicle during the roll phase is a type of debris that often remains present on the off-road surface for a significant period of time following the accident. Thus, it can be discovered and located by investigators present at an accident site long after the accident occurred. This is consistent with

FIGURE 2.14 Close-up of a glass sample collected from Area 1, site of Test #6 from Asay [2], 6 years and several months after the test.

our experience visiting the site of this test. Perl observed, "some simple analysis may be applied to narrow down the candidate windows; for example, the rear windows of most vehicles contain metal strips for window defrosting, and often the window tinting is different for various positions. The specified nominal thickness of the tempered glass in the various positions often is different. For example, in an SUV the rear quarter glass may be specified to be 4.0 millimeters (mm) while the adjustable door windows may be 3.5 mm."

Przybyla [7] studied the longevity of glass debris over a period of 2 years. He placed glass debris at three locations in the United States where they would be subjected to the influence of slope, weather, mowing operations, and other factors that could result in movement of the debris field. Each area of glass debris was periodically mapped to determine the degree to which the glass debris had moved or changed in size. Przybyla documented expansion and movement of each glass debris field over the study period, but these were small. He ultimately concluded that "roadside glass debris fields remain generally static over time when subjected to natural weather patterns, normal landscape growth, and roadside mowing."

In addition to the glass deposits identified during our inspection, the locations of individual bushes that were visible in the original photographs from the test and that were still visible during this inspection made the rest position of the vehicle identifiable. On the photograph of Figure 2.15 (taken immediately after the test with the vehicle still at rest), four specific bushes and an open dirt area around those bushes have been identified. These same four bushes and the open dirt area have also been identified in the image of Figure 2.16, which is one frame of a video taken from a UAV at the time our inspection.

Figure 2.17 is an aerial photograph from our inspection, again taken with a camera attached to a UAV. In this aerial, we have identified the four bushes from Figure 2.15, along with two additional bushes, visible in the posttest photograph and also during our inspection. These figures demonstrate that from using the vegetation still in place 6 years and several months after the test, the rest position of the vehicle could be determined.

Our ability to locate evidence at this site was likely, to some degree, unique to this particular location, with its arid desert conditions. Over the course of more than 6 years,

FIGURE 2.15 Photograph showing bushes surrounding the vehicle rest position.

FIGURE 2.16 One frame of video from UAV footage showing the bushes of Figure 2.14, 6 years and several months after the test.

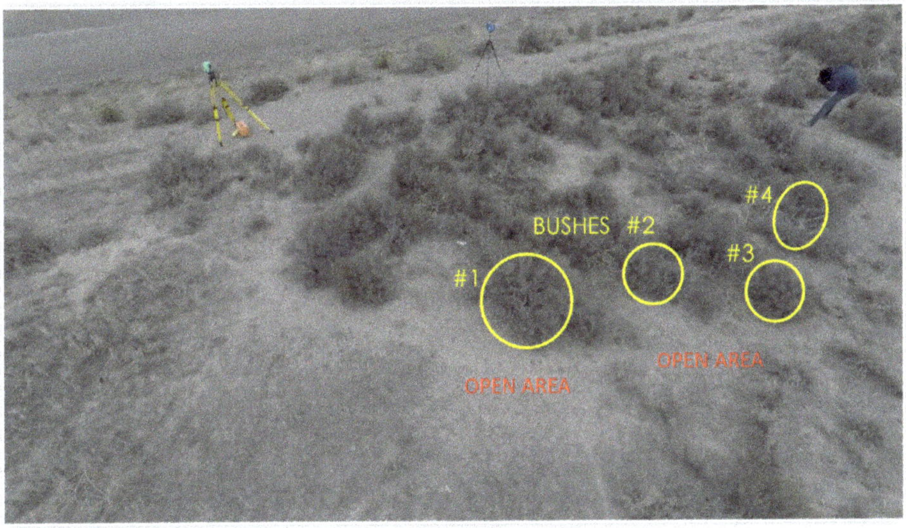

FIGURE 2.17 One frame of video from UAV footage showing the bushes of Figures 2.14 and 2.16, 6 years and several months after the test.

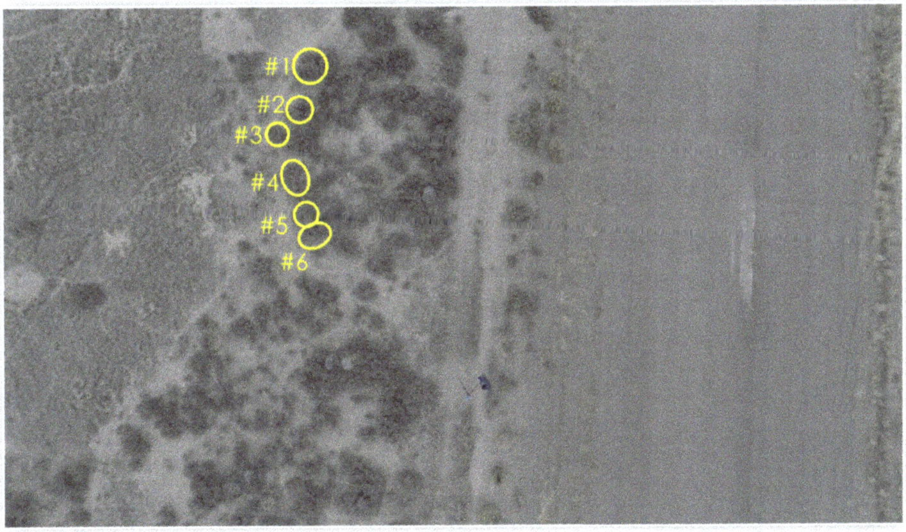

it is possible that more significant changes would occur to vegetation in a more humid or moist environment. In areas of the United States with more significant annual rainfall and a warmer climate, it is possible that glass deposits would be obscured by growing vegetation or covered by foliage. It is also possible that other sites would exhibit less significant roadway changes even over a 6-year timeframe. Regardless, this case study shows that when investigating and reconstructing a rollover crash, even more 6 years after the crash occurred, a site inspection could provide useful and pertinent evidence for the reconstruction.

The following figures depict evidence deposited on the roadway and ground during several real-world rollover crashes. Figure 2.18 contains three photographs of

FIGURE 2.18 Photographs of striated yaw marks from a rollover crash.

a set of yaw marks deposited on a roadway during a loss of control that preceded a rollover. These yaw marks exhibit striations, which can sometimes be used to infer driver steering, braking, or acceleration inputs (see Chapter 6). If possible, the angle of these striations can be documented during a scene inspection. Figure 2.19 is another yaw mark from a real-world rollover. This yaw mark again exhibits striations which are made more visible by the fact that the mark traverses a section of roadway striping [see Tandy et al. [8], in *Collision*, for a detailed discussion of tire mark types]. Figure 2.20 depicts an example of glass debris fields from a rollover. The photographs show four distinct areas of glass deposits - two that are clear and two that are tinted. Other examples of glass debris are depicted in the photographs of Figures 2.21 through 2.23. The last of these depicts a piece of glass with a decal that could likely be used to identify which window of the vehicle this glass deposit came from. It is worth noting also that the side mirror glass of a vehicle can provide insight as to which side of the vehicle landed at a particular location. A distinguishing factor between the driver- and passenger-side mirrors is the laser-etched wording "OBJECTS IN MIRROR ARE CLOSER THAN THEY APPEAR" on the lower portion of all passenger-side mirrors.

FIGURE 2.19 Photograph of a yaw mark with striations from a rollover crash.

When vehicles roll on-road or across a paved surface, the wheels often gouge the pavement. An example of this gouging is shown in Figure 2.24. In this instance, nearly the entire circumference of the rim made contact with the pavement. It is likely that the vehicle struck the surface in either a ¼ or ¾ roll orientation in order to deposit this gouge. Other nearby evidence may distinguish between

FIGURE 2.20 Photograph of glass debris from a rollover crash (Example 1).

FIGURE 2.21 Photograph of glass debris from a rollover crash (Example 2).

¼ or ¾ roll positions. When only a portion of the rim makes contact, an arc-shaped gouge is typical. The direction of the arc can help determine whether the right-side or left-side rims made contact. Many times, accompanying the arc-shaped gouge, there will also be a smeared tire mark located at the base of the arc from where the tire has folded under exposing the rim upon landing. Arc-shaped gouges that open away from the point of rest are typically deposited from the trailing-side rims. Arc-shaped gouges that open towards the point of rest are deposited from the leading-side rims.

FIGURE 2.22 Photograph of glass debris from a rollover crash (Example 3).

FIGURE 2.23 Photograph of glass debris from a rollover crash (Example 4).

FIGURE 2.24 Gouge mark deposited from entire rim circumference.

Vehicle Evidence

When a body panel of a vehicle contacts the ground during a rollover, scratch marks are often created on the panel. Figure 2.25 depicts the right rear of the Ford sport utility vehicle after the dolly rollover reported by Rose [9]. In this photograph, scratch marks are visible on the rear-side window glass, the rear quarter panel, and the wheel well trim. Figure 2.26 depicts similar scratch marks on the hood of a vehicle that was involved in a real-world rollover. The vehicle depicted in Figure 2.26 rolled over multiple times and exhibits families of overlapping scratch marks. These scratch marks can be used for multiple purposes.

FIGURE 2.25 Scratch marks on a vehicle after a dolly rollover test.

FIGURE 2.26 Overlapping scratch marks on the hood of a vehicle.

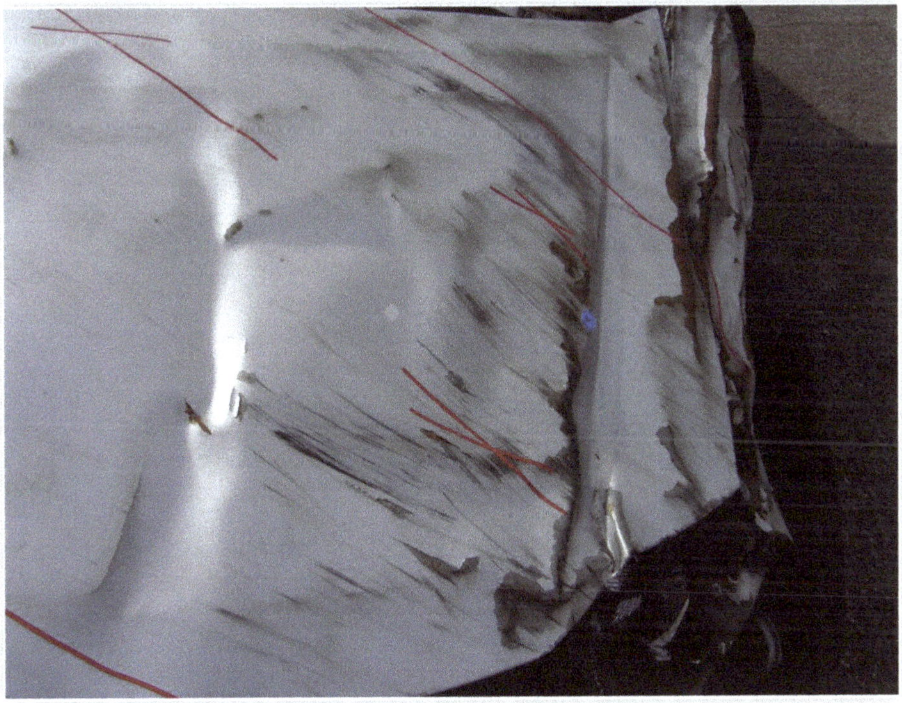

For instance, in combination with other evidence and analysis, they can be used to determine the yaw orientation of the vehicle at the point during the roll where the scratches were created. Also, on body panels where there are multiple families of scratches with different orientations, overlapping scratches can be used as an estimate as to how many times that part of the vehicle was in contact with the ground during the rollover [10]. This is an indication of the minimum number of rolls experienced by the vehicle during the crash - minimum because a vehicle can complete a full roll while airborne [2, 11]. Additionally, the surface the vehicle rolls on may not deposit scratch marks on the body panels, either because the surface is soft or because the vehicle only touches down lightly or briefly [10].

Orlowski [10] observed that "the key to reconstructing these [scratch] patterns is to learn the direction of each pattern of scratches and which came first." He noted that, to determine the order in which the scratches were deposited, "it is important to find areas where the multiple scratches are of similar depths. The best indicator of which scratch came first can come from careful magnified examination of the intersection of scratches. Where one scratch crosses another, careful study of the intersection can reveal metal flow or disturbed edges that show which was first. Another clue to the sequence of collisions is the presence of scratches either within folds of sheet metal or on ridges of wrinkled sheet metal. If the sheet metal deformation clearly came from a different blow than the impact which caused the scratches, the sequence of those two impacts is defined."

Turriff [12] examined the characteristics of scratches that would indicate the scratch direction. He reported results from four dolly rollover tests run with 2001 Chevrolet Trackers. Each test was run with the driver's side of the vehicle leading and the dolly traveling a nominal speed of 35 mph (56 kph) prior to the sudden deceleration that initiated the separation of the vehicle from the dolly and the rollover. All four vehicles rolled between 1 and 1½ times, initially on a smooth concrete surface when they began to roll and then traveling onto a level dirt surface where they continued to roll. A small amount of the dirt (a sandy aggregate mixture) was reportedly present on the concrete surface. Turriff found that "the most evident indicators of scratch direction, and thus roll direction, were accumulations of abraded material found at the termination points of scratch-damaged areas. Abrasive wear mechanisms caused local plastic deformation patterns that were evident on painted sheet metal surfaces as well as plastic trim pieces. In both cases, abraded material is plowed towards and accumulates at the end of the scratches... At the end of the scratch, a raised berm of deformed, plowed metal and partially detached metal chips or fragments can be found." This technique can also be applied to plastic trim and molding that tend to "flow" in the direction they are being displaced. Interestingly, another valuable inference from the general scratch direction of families of scratches is the fact that they point to the side of the vehicle that is leading.

Orlowski noted a potential problem with scratch pattern analysis, stating that "the shape of the deformed vehicle at the time those scratches were made must be known. Slight changes in the exterior shape of the car can appear to indicate great changes in the yaw angle at the time of the impact based upon the scratches..." Turriff also noted this difficulty with scratch analysis, observing on the one hand that "since a specific vehicle surface may not touch down during every roll, the number of distinct scratch groups reflects only the *minimum* number of vehicle rolls needed to cause the observed scratch groups" but also noting that "the difficulty in using scratch groups to determine the number of rolls lies in identifying distinct groups." Discussing one of his tests, Turriff observed, "despite rolling only once...there were scratches on the roof...that were similarly though not identically aligned...These scratches all occurred during one prolonged period of sliding while the vehicle rotated about its vertical axis. Roof crush that occurs during sliding contact can also account for variations in apparent scratch angle within a single scratch group. Scratch orientation varies because the scratched surfaces are progressively deformed, crushed, buckled or otherwise displaced relative to the rest of the vehicle as new scratches are being made...As a result of yaw-, sliding- and crush-induced variations in scratch orientation,

FIGURE 2.27 Photographs showing the window glass on the Explorer after the test.

care must be taken not to interpret these variations in scratch orientation as evidence for separate groups of scratches that occurred during separate vehicle rolls."

Bready [11] discussed the effect of roll rate on scratch pattern direction. He noted that "evidence from the scene and vehicle...can sometimes appear inconsistent, because they suggest a substantially different vehicle orientation at a particular contact point." Bready suggests that this inconsistency is due to the incorrect assumption that the scratch direction can simply be aligned to the over-the-ground velocity direction of the vehicle to determine the vehicle orientation. "If a vehicle were simply sliding on the ground without any rolling motion, then the resulting scrapes would be parallel to, and opposite in direction of, the vehicle horizontal velocity... If a vehicle was only rotating, and not translating over the ground, then the resulting scrapes would be parallel to, and opposite in direction of, the roll motion of the vehicle, and its associated peripheral velocity. However, there are generally both roll and translational motions in a rollover accident. The contribution of these two components determines the actual velocity of the vehicle contact point, relative to the ground, and will determine the angle and direction of the scrape marks left on the vehicle."

Returning to Asay's Ford Explorer test, Figure 2.27 shows the vehicle at rest following the test. These photographs reveal that during this rollover, most of the window glass on the driver's side of the vehicle broke and separated from the vehicle while the passenger's side window glass remained intact. The rear window glass also broke and separated from the vehicle. Clearly, identifying which windows did not break during a crash helps to narrow down from which window a glass deposit came. Figure 2.28 shows portions of another vehicle that was involved in a real-world rollover. Examination of this photograph reveals glass fragments that remain in and around the window openings on this vehicle. Figure 2.29 shows a measurement being taken of the thickness of some of

FIGURE 2.28 Photograph showing glass debris remaining on a vehicle after a real-world rollover.

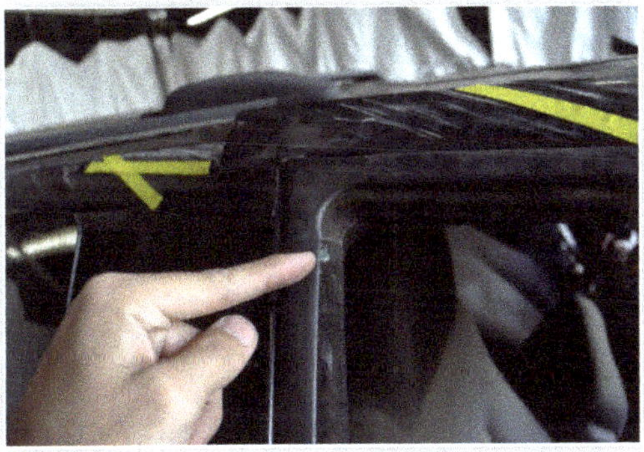

FIGURE 2.29 Photograph showing measurement of the thickness of glass debris remaining on a vehicle after a real-world rollover.

this glass. Perl [6] conducted a survey of 108 panes of tempered glass from 18 vehicles. He concluded that the difference in thickness from window to window on a vehicle would often be sufficient to distinguish one window from another. A reconstructionist can map the thicknesses of a vehicle's windows using glass fragments stuck or remaining in the window openings and then compare these measurements to measurements taken from glass debris fields at the accident site.

When a vehicle rolls over yellow fog lines, yellow paint can be deposited on the vehicle. An example of this is depicted in Figure 2.30. This evidence, along with scrapes on the roadway, assists in determining vehicle orientation during the roll sequence. This can occur with lane lines as well, although the white paint can be more difficult to identify on the vehicle.

Vehicle rims and tires often display evidence of ground contact. The rim in Figure 2.31 was bent and imbedded with asphalt as result of a ground strike. The damaged rim circumference can be matched to similarly sized gouges on the roadway to determine where the impact occurred. It is common for tires to be pushed up into the wheel well during significant ground impacts, leaving tire impressions. Figure 2.32 is an example of this type of wheel well evidence.

An additional aspect of the vehicle damage that may be relevant is damage to the doors of the vehicle. Bandstra [13] reported a series of rollover tests where passenger vehicles were rolled down a 32° dirt embankment. For these tests, the striker pin, or striker plate, was removed on the driver's side door of each test vehicle in order to ensure this door would open in each of the tests. This resulted in a situation for each test where

FIGURE 2.30 Yellow paint transferred to the vehicle from the roadway fog line.

FIGURE 2.31 Rim damage from striking the paved road.

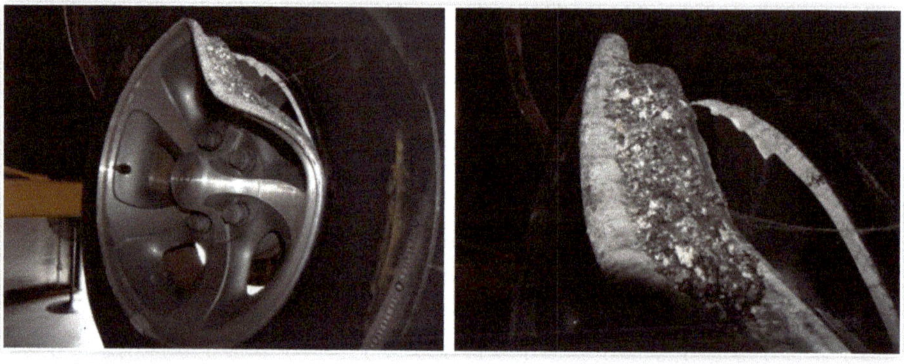

the driver's door would open during the test but the passenger's door would not. Based on this testing, Bandstra compared the damage to the doors that were open during the rollover to the doors that were not open. He noted the following characteristics of the damage to the doors that were open during the rollover: (1) the door no longer fits the opening of the vehicle body, "indicating nonsynchronous damage to the door and surrounding structures"; (2) the door is misaligned relative to the vehicle and the other door when viewed from the front or rear with both doors open; (3) the trailing edge of the front fender and leading edge of the door show contact damage from over-rotation of the hinges; (4) the fender well opening or wheel rim may leave an impression on the outer door skin.

FIGURE 2.32 Wheel well damage from a ground strike.

In some instances, accident reconstructionists may be asked to evaluate the manner in which vehicle occupants interacted with the interior of a vehicle during a rollover or the manner and timing of an occupant ejection. In these instances, the reconstructionist needs to document the interior of the vehicle. Raphael et al. [14], for instance, discussed physical evidence that would be present on a seat belt system if a seatbelt came unbuckled during the rollover. Specifically, Raphael noted that, in such instances, "the seatbelt webbing entangled the outboard upper extremity [of the occupant] and produced severe injuries. When the occupant was ejected from the vehicle, the seatbelt was also loaded so that significant marks were produced on the exposed webbing fabric and/or the vehicle trim."

Scene and Vehicle Documentation Checklists

The discussion of the previous sections suggests various pieces of evidence from a rollover crash that can be documented during a physical or photographic inspection of a crash site or vehicle. This section summarizes the evidence discussed in the previous sections by formulating them in terms of a checklist for the site inspection and for the vehicle inspection, assuming the reconstructionist is able to physically inspect both. In addition to that, these checklists add additional details that the reconstructionists may want to document. These lists are suggestions that are intended to help the reconstructionist formulate their own checklist for any particular case.

Site Inspection Checklist

The evidence that is documented during a site inspection depends, in part, on what issues you are analyzing and on how much time has elapsed between the crash and your inspection. For the sake of completeness, the list that follows assumes that your site documentation is taking place before any evidence has deteriorated. The items on this list are simply suggestions. Your checklist for any particular case may have more or less items, depending on the issues you are analyzing.

- **Take overall photographs of the accident site.** Typically, this involves starting at the beginning of the physical evidence (initial tire marks from the loss of control, for instance) and walking along the full trajectory of the vehicle, taking photographs along the way. Aerial photography taken from a UAV can also be useful for overall documentation of the accident site.

- **Map the physical evidence from the crash.** Mapping will typically be accomplished with a total station or a laser scanner, though image-based scanning could also be used.
  - ☐ Tire marks
  - ☐ Furrows
  - ☐ Gouges and scrapes
  - ☐ Glass deposits and thickness
  - ☐ Mirrors or deposits of mirror glass
  - ☐ Ground impact marks
- **Map the site geometry.**
  - ☐ Roadway striping
  - ☐ Terrain
  - ☐ Signage
  - ☐ Vegetation and trees
- **Document site characteristics.**
  - ☐ Speed limit.
  - ☐ Latitude and longitude.
  - ☐ Signage.
  - ☐ Note if any surveillance cameras are present.

Vehicle Inspection Checklist

The evidence that is documented during a vehicle inspection will also depend on what issues you are analyzing. The items on this list are simply suggestions. Your checklist for any particular case may have more or less items, depending on the issues you are analyzing.

- **Take overall photographs of the vehicle.** Typically, a set of overall photographs would consist of 8 photographs - one taken from each side of the vehicle and one take from each corner of the vehicle.
- **Document the vehicle identification number (VIN) and placards.**
- **Document the damage to the vehicle,** taking photographs with a wide range of zoom levels. While documenting damage to the vehicle, determine if the damaged condition of the vehicle was altered during the efforts of emergency personnel to extract occupants after the crash.
- **Document the tires and wheels of the vehicle.**
  - ☐ Write down or photograph the tire sidewall information from all four tires.
  - ☐ Measure the tread depth of each tire.
  - ☐ Measure the tire pressure of each tire.
  - ☐ Photographically document damage to any of the wheels.
  - ☐ Photographically document any evidence of contact between a tire and its wheel well.
  - ☐ Photographically document any asphalt or soil deposits in and around the wheels and tires.

- **Document the windows and mirrors of the vehicle.**
  - ☐ Identify which windows are broken out and which are still intact.
  - ☐ Identify which windows were up and which were down at the time of the crash.
  - ☐ For windows that are broken, take thickness measurements of glass fragments still present on the vehicle.
  - ☐ Document which windows are tinted and which are not.
  - ☐ Document which windows are constructed of tempered glass and which are constructed of laminated glass.
- **Document the body panels of the vehicle.**
  - ☐ Identify scratches on the body panels. Use tape to document their direction. This tape helps make the scratch directions more apparent in the photographs you take.
  - ☐ Some investigators use a different color tape for each scratch family on the vehicle. At times, this may be problematic since it may not be apparent, simply from visual inspection of the vehicle, which scratch marks on the various body panels are related. It is recommended that a single color of tape be used for scratch marks that cannot be definitively grouped so that the process of separating these into particular families can also account for the site evidence and characteristics of the speed and roll velocity curves that result from analysis of the evidence.
  - ☐ The order of overlapping scratch marks on the vehicle can often be determined in the field. If this is the case, one could apply the tape to the vehicle along the scratch directions in the same order the scratch marks were deposited during the accident.
  - ☐ Directional arrows can be added to the tape, but it is recommended to add them only in areas where there is definitive evidence of material flow patterns. These can aid the reconstructionist in defining the roll direction.
  - ☐ After you have marked the scratch mark directions, take another set of overall photographs of the vehicle.
  - ☐ Take close-up photographs of the scratch marks on the vehicle, as well. Progress through these photographs in such a way that it is clear what part of the vehicle is being photographed. For example, to photograph scratches on the passenger-side door, begin with a zoomed out photograph of the passenger side of the vehicle and progressively zoom in to a close-up of the scratches. The progression from photograph to photograph makes it clear what part of the vehicle is being photographed in the zoomed in views.
- **Document the contents of the vehicle.**
  - ☐ Oftentimes, vehicle components that separated from the vehicle during the crash will be placed inside of the vehicle when the vehicle is removed from the scene. Removing these components from the vehicle and documenting damage and scratch marks on them can yield additional evidence for the reconstruction.
  - ☐ Documentation of the vehicle contents may also reveal what cargo was in the vehicle at the time of the crash.
- **Measure the postcrash damage to the vehicle.** Oftentimes, this will be accomplished with a laser scanner. However, other techniques are also available, such as image-based scanning.

CHAPTER 2

- **Document any changes to the vehicle since it was removed from the accident scene.**
  - ☐ Observe and list any damage to the vehicle or changes that have occurred since the crash, based on a comparison with photographs taken at the accident site. This may include damage sustained by removing the vehicle from the site of the accident, transporting, or vehicle storage.
  - ☐ Document any unusual prints or marks observed on the vehicle.
- **Image the data from any event data recorders on the vehicle.**

Photogrammetry

Reconstructionists are often asked to analyze a crash long after it has occurred. This usually means that some of the physical evidence that was present at the accident scene is no longer physically at the site. Sometimes, the involved vehicle is no longer available for physical inspection. In these instances, reconstructionists can apply techniques of photogrammetry to map out the evidence, as long as the evidence was photographed to begin with. The term *photogrammetry* refers to techniques used for "obtaining reliable information about physical objects and the environment through processes of recording, measuring, and interpreting photographic images" [15]. Tumbas defined photogrammetry as "a scientific method for determining the dimensions of an object by measuring a photographic image of the object and transforming measurements of the image to the actual surface" [16]. The term *close-range photogrammetry* refers to applications of photogrammetry where the camera-to-object distances are on the order of "tens and hundreds of yards as opposed to the tens and hundreds of miles associated with aerial uses" [17].

Close-range photogrammetry has long been used for accident scene documentation, and Niederer [18] documents its use by the city police of Zurich beginning as far back as 1934. Pepe [17] described the origins of photogrammetry, stating that "photogrammetry has its beginnings with Leonardo da Vinci and his investigations of optics and the geometry of projection. Da Vinci's contemporary Albrect Dürer developed the laws of perspective which were the basis for Henry Lambert's classic treatise 'The Free Perspective' written in 1759. It outlined the principles which today comprise the science of [sic] known as photogrammetry. When, in 1837, Louis Daguerre presented to the French Academy of the Arts and Sciences the first tangible photographic results based on emulsion films, most of the components were in place."

As they are applied to accident reconstruction, photogrammetric techniques include single-image and multiple-image methods. Some of these techniques are implemented graphically and others are implemented mathematically [19]. Some techniques use calibrated, or metric, cameras for which key dimensions and characteristics of the camera are known and fixed, while others use uncalibrated, or nonmetric, cameras. Calibrated cameras provide more up-front information for the analysis, and thus, techniques using these require fewer known scene dimensions [16]. A number of photogrammetric techniques are discussed in the literature and used in the accident reconstruction community. The discussion that follows describes a number of these techniques.

Rectification

One technique commonly used in accident reconstruction is referred to as *image rectification*. In this technique, a photograph is projected onto a surface representing a portion of the accident scene. The visual effect of this is that the image appears flattened and stretched, such that the lines and shapes in the image can be aligned to a scaled scene diagram from a top down perspective. Niederer et al. [18] discussed the use of "simple perspective rectification" for determining the location and dimensions of physical evidence depicted in accident scene photographs. The process described by Niederer assumed the area being analyzed could be approximated as a single plane. Niederer compared results from single-image rectification to mapping of the same accident scene using stereophotogrammetry and reported that "the maximal deviation of the map obtained by single-view photogrammetry from the reference is less than 20 cm within the region of interest." He attributed the error to non-planarity of the accident scene, insufficient image contrast (making point identification more difficult), and lens distortion.

Neale [20] has described and validated image rectification techniques (projection mapping) that do not assume the mapping surface is planar. He has applied these techniques to single images and to video footage. Neale evaluated the accuracy of three-dimensional projection mapping by using it to measure the length of several mock tire marks and comparing the measured length to the length obtained with a digital survey. The surveyed lengths of the tire marks varied between 8 and 109 ft. The projection mapped tire mark lengths differed from the survey by between 1 and 3.5 in.

Camera Matching

Another technique, referred to as *camera matching*[9] in the more recent literature, involves reconstructing the location from which a photograph was taken and determining the field of view of the camera that took it. Once the camera location and field of view are obtained, objects (physical evidence) within the photograph can be located. This graphical technique, which is often used for diagramming scene or vehicle evidence, involves the following steps:

(1) The reconstructionist selects one or more photographs for analysis. Often, these photographs show evidence at an accident scene (tire marks or gouges, for instance) or damage to a vehicle. Photogrammetry is often employed when this evidence no longer exists at the scene or the vehicle is not available for physical inspection. Photogrammetry allows this evidence to be diagrammed through analysis of photographs even though it cannot be physically measured. In order for a photograph to be analyzed, it needs to show some objects or geometry that still exist and can be physically documented. For analysis of scene evidence, such geometry would typically be roadway striping, curbs, signs, or trees. For analysis of vehicle damage, such geometry is often undamaged portions of the vehicle that can be documented through examination of an exemplar vehicle. As an example, Figure 2.33 depicts a scene photograph chosen for analysis. Specifically, this analysis will determine the location of a set of tire marks and the rest position of the vehicle.

(2) After selecting photographs for analysis, the reconstructionist makes note of objects or geometry shown in the photographs that still exist at the scene or parts of the vehicle that can be documented on an exemplar vehicle. The reconstructionist then physically documents this geometry at the scene or on an exemplar vehicle using a total station or laser scanner. In the example of Figure 2.33, the roadway striping, the trees, the fences, and the signs were surveyed.

[9] Baker [15] refers to the camera matching technique as "computer reverse projection." This terminology has the advantage of showing the technique's connection to a prior technique discussed in the literature, often referred to as "camera reverse projection." That said, these authors prefer the term "camera matching" because it more clearly communicates to a lay audience (a jury, for instance) what the technique is actually doing.

FIGURE 2.33 Photograph chosen for example.

(3) After the scene or exemplar vehicle mapping data has been processed, the reconstructionist uses a computer modeling software package to create a virtual camera and to view the survey or scan data from a perspective that is visually similar to that shown in the photograph that is being analyzed.

(4) Lens distortion is removed from the photograph that is to be analyzed. All camera lenses produce some level of distortion in the resulting image, and this distortion can potentially introduce errors into a photogrammetric process if it is not removed. A 2011 study by Neale et al. [21], examined the lens distortion caused by 35 different makes and models of camera lenses and then discussed methods and software available for removing the distortion from the resulting images. Neale noted that for a particular focal length, two lenses of the same make and model will produce the same distortion. Lens distortion occurs in two basic forms - barreling and pin cushioning. Barreling is distortion that squashes the edge of the image, whereas pin cushioning is distortion that stretches the edge of the image. Generally, barreling occurs from wide focal lengths and pin cushioning occurs from zoom focal lengths, although lenses can also produce a mix of barreling and pin cushioning.

(5) The corrected photograph or image is then imported into the modeling software and is designated as a background image for the virtual camera.

(6) The analyst then makes adjustments to the location, focal length, and viewing plane of the computer-modeled camera until an overlay is achieved between the survey or scan data and the scene or vehicle geometry shown in the photograph. Once a match is obtained, the analyst has reconstructed the location, focal length, and viewing plane of the camera used to take the original photograph. In Figure 2.34, the survey data has been aligned to the geometry depicted in the photograph.

(7) Once the camera location and characteristics are obtained, the evidence visible in the photograph can be traced or modeled such that it also overlays what is visible in the photograph. In Figure 2.35, the tire mark evidence and rest position of the vehicle have been located. In Figure 2.36 the photograph has been removed, leaving the scene model, physical evidence, and rest position.

The camera matching technique is a single-image photogrammetric method, in the sense that a single image is sufficient for its application. However, multiple photographs depicting the same evidence from different vantage points can be analyzed, and when

FIGURE 2.34 Survey and scan data aligned to the photograph.

FIGURE 2.35 Tracing of evidence once camera location and characteristics have been determined.

FIGURE 2.36 The photograph has been removed leaving the scene model and evidence.

the results are averaged, this may increase the accuracy of the technique. Campbell and Friedrich first described the camera matching technique in the accident reconstruction literature in 1993. Using a mock accident scene, they compared the positions of points obtained with the technique to the actual positions of these points. They reported a minimum position error of 0.1 ft and a maximum position error of 1.74 ft. In a 1999 study, Massa [22] demonstrated that the technique could be applied to analyze and critique an animation. In a 2001 study, Fenton [23] described and tested the camera matching technique for obtaining dimension from another mock accident scene. He compared dimensions obtained from camera matching with the same dimensions obtained with a digital survey and reported an average error of around 2%, with a range of errors between 0.48% and 8.23%. The dimensions ranged from approximately 5 ft up to approximately 140 ft.

The camera matching technique can also be applied to frames from a video. Chou et al. [24] reported application of the technique for tracking the motion (position, roll angle, and roll velocity) of a vehicle from video of a dolly rollover crash test. The tracked roll angle and calculated roll velocity were compared to the roll velocity measured with a sensor and the roll angle obtained from integration of the sensor data. Chou reported excellent agreement between the roll angles determined from camera matching and those determined from the sensor data. There were discrepancies between the roll velocities obtained from camera matching and those obtained from the sensor, though the trend of the sensor data was generally matched with the camera matching technique.

Rose et al. [9] reported additional application of the camera matching technique to analysis of a dolly rollover crash test, attempting to improve on the results from Chou [24] through more extensive knowledge of the locations and characteristics of the cameras that recorded the test and the geometry of the test facility. In this study, a digital survey was conducted to locate the cameras, to document the test facility geometry, and to locate the targets on the vehicle. This reduced the number of unknowns in the camera matching. In addition to that, knowing the characteristics of the cameras made it easier for lens distortion to be removed from the video prior to the analysis.

Rose reported the motion of the vehicle for a 2-s segment of the test and compared the roll velocity results to sensor data. The resulting roll velocities from the camera matching were generally bracketed by the signals from the two roll velocity sensors on the vehicle. In other words, the discrepancy between the roll velocity from the camera matching and the roll velocities from either sensor were less than the discrepancy between the two sensors. Rose also reported a comparison of 2 separate analysts using the camera matching technique on the same 2-s segment. Around 96% of the time, the vehicle position obtained by the two analysts was different by less than an inch. This propagated to an uncertainty in the calculated over the ground and vertical speeds of 1 mph with a confidence of 96%. For the vehicle roll angle, the difference between the two analysts was less than 1°, approximately 85% of the time. For an analysis time step of 40 ms, this propagated to an uncertainty in the roll velocity of approximately 18 deg/s with a confidence of 85%.

Coleman et al. [25] examined the accuracy of the camera matching technique when implemented with laser scan data rather than a digital survey. They tested the accuracy of the camera matching technique by using it to locate evidence placed at a mock crash site and then comparing the obtained evidence locations to the actual locations of the evidence. They noted that "in many places the evidence location error was near zero." Their worst-case error for the camera matching technique was about 13 centimeters (5.1 in.).

Camera Reverse Projection

The camera matching technique is essentially a digital implementation of a technique the earlier literature referred to as *camera reverse projection*. According to Tumbas [16], this term refers to the fact that once the location and characteristics of the camera are known, points from the photograph can be projected onto the scene. This technique followed a similar method to that described for camera matching, but the implementation did not use computer modeling programs or virtual cameras. Instead, the reconstructionist would overlay a transparency of the original photographic print and trace geometry that was visible in the photograph that would still be visible at the site. A small version of this transparency would then be created and inserted into the viewfinder of what the Northwestern University Traffic Collision Investigation manual referred to as a "dummy" camera. While at the accident site, the reconstructionist would then change their viewing location and the camera focal length until the features traced onto the transparency overlaid on the corresponding features at the accident site. Once the viewing location, viewing plane, and focal length for the original photograph were reconstructed, the reconstructionist could then locate physical evidence that was no longer visible at the site. One limitation in this technique is that the analyst would need to use a camera conducive to inserting a small transparency into the viewfinder. This would usually mean that the analyst would be using a different camera with different internal geometry than the camera that actually took the photograph. Another limitation is that the analysist would have to stand in the same place as the original photographer for an extended period of time. If the photograph to be analyzed was taken from the middle of a lane on the interstate, for example, the analysis may not be possible from a safety standpoint. Digital implementation of the camera matching technique does not suffer from these limitations, since the analyst has greater control over the characteristics of the computer-modeled camera and does not have to stand in the road. Another limitation of the older implementations of camera reverse projection was that there was no real way to account for or eliminate lens distortion from the photographic image. Digital implementation of the camera matching technique is able to easily correct or eliminate distortion from many photographic images.

Despite these limitations, many reconstructionists successfully and accurately implemented the camera reverse projection technique. Breen and Anderson [26] described an implementation of reverse camera projection to determine the postaccident shape of a damaged windshield. Woolley et al. [27] applied camera reverse projection with two photographs to quantify vehicle crush. He reported average errors of less than 1 in., noting though that this level of accuracy "depends on the experience, patience, and care taken in all phases of the process. The more times camera reverse-projection is applied, either to a damaged vehicle or to a scene, the greater will become the skill of the analyst." This statement also applies to a digitally implemented camera matching process. However, what Woolley has left unsaid is the important fact that the application of the reverse camera projection and camera matching techniques typically result in a graphical output that allows others to judge the quality of the camera match.

Husher et al. [19] applied the camera reverse projection technique to a mock accident scene and reported that it "was able to locate the subject skid to within 20 cm (8 inches) for the least favorable camera view, and within 5 cm (2 inches) for the more favorable views. The selection of the camera view determines the accuracy of this method." They also stated that "when the results of the three [camera reverse projection] views were averaged, the determined position was almost on top of the true position for both the skid and the vehicle." This means that, though camera reverse projection is methodologically a single-image technique, applying it to multiple images and averaging the results improves the accuracy. In this study, camera reverse projection was also applied to determine vehicle crush, and the authors reported that the "longitudinal crush determined was within 2.5 cm (1 inch) of the measured value."

Husher also described an implementation of the camera reverse projection technique that involved "real time analysis with video equipment…Two video cameras view the original photo and the exemplar vehicle (or scene) simultaneously and a video mixer is used to overlay the two separate views to provide the alignment information…." Husher applied this method to determining vehicle crush and reported that "accuracy was within 2.5 cm (1 inch) of the crush." Main and Knopf [28] describe an implementation of the camera reverse projection technique using physical scale models of the accident site and vehicles. Though computer implementation of the camera matching technique began appearing in the literature as far back as 1993 [29], the technique described by Main and Knopf is methodologically a direct precursor to the camera matching technique.

Smith [30] and Tumbas [16] noted that reverse camera projection can be accomplished analytically with two photographs. Tumbas stated this would be accomplished by "surveying selected points at the site, constructing a mathematical model of the site surfaces, mathematically locating the camera, mathematically projecting the points of interest onto the model and determining the location of the intersecting rays with the model's surface." Referring to this technique as *analytical reverse projection*, Smith described the concept of this method as follows: "This method is based on the fact that once the original position of the camera lens center, the distance from the lens center to the picture-plane (photograph), and the orientation of the photograph relative to the scene are all determined, then a vector passing from that established lens center point through any point of interest in the two-dimensional picture-plane defines the ray that passes through that same point of interest in the actual three-dimensional scene. If two or more photographs, taken from different locations, are available showing the same point of interest, then the location of that point in the scene can be determined by simply calculating the point where the two rays (on from each photograph) intersect." Tumbas reported typical accuracies for locating points with reverse camera projection of less than a foot, with maximum errors around 2 ft. For analytical reverse projection, he reported typical accuracies of around a half foot, with maximum errors of less than a foot. Tumbas notes that both techniques resulted in better accuracy than methods that assume the surface containing the evidence is flat. Reverse camera projection does not require such an assumption.

PhotoModeler

PhotoModeler is a photogrammetry software package commonly used by accident reconstructionists. The literature contains extensive coverage of this software and its capabilities. In 1997, Fenton and Kerr [31] described the use of PhotoModeler for implementing camera reverse projection and rectification with accident scene photographs. They presented a case study for which they compared several scene measurements made at the scene with the same measurements obtained using PhotoModeler. They reported that the measurements obtained from PhotoModeler were within 1% of the field measurements. The measured distances varied between 10 and 17 ft. In 1999, Fenton et al. [32] described the use of this software for determining vehicle crush, indicating that under typical circumstances, it would yield an accuracy of ±1 in. in the crush measurements. PhotoModeler software could be used to quantify the roof deformation exhibited by a vehicle after a rollover. Whereas the camera matching technique is a single-image method, the use of PhotoModeler to determine vehicle crush requires at least two photographs showing the same portion of the vehicle. The vantage point of these photographs would ideally be separated by a large angle of around 90°.

Also in 1999, Switzer and Candrlic [33] published a study of the factors affecting the accuracy of PhotoModeler. They noted that "PhotoModeler solves for the intersection location of rays from the same points in each image by solving the collinearity equations for all the selected points and iterating the camera location and point locations to obtain a minimum least square error for the set. PhotoModeler also solves for the camera focal length, principal point, and format aspect ratio." Switzer and Candrlic reported that the number of control points had a significant effect on the accuracy, as did the amount of image cropping. Image cropping reduced the accuracy of the analysis.

Peck used PhotoModeler to measure three targeted vehicles [34]. To optimize the method, he utilized a field-calibrated DSLR camera equipped with a fixed 20 mm lens, retroreflective targets, and a machined scale bar. Peck utilized 8 photographs per vehicle. He compared the measurements he took with PhotoModeler to measurements taken with a total station and found that "the average residual between corresponding photogrammetry and total station points was 1.7 mm (N = 258, SD = 0.8 mm) with a 95% confidence limit of 3.1 mm." He also measured one of the vehicles with a FaroArm. He found that "the average residual between corresponding photogrammetry and FaroArm points was 1.2 mm (N = 83, SD = 0.56 mm) with a 95% confidence limit of 2.1 mm."

References

1. Cofone, J.N., *Rollover Crashes of Motor Vehicles and Heavy Trucks*, (Jacksonville, FL, Institute of Police Technology and Management, 2002), ISBN 978-1-884566-50-9.
2. Asay, A. and Woolley, R., "Rollover Testing of Sport Utility Vehicles (SUVs) on an Actual Highway," SAE Technical Paper 2010-01-0521, 2010, doi:10.4271/2010-01-0521.
3. Asay, A. and Woolley, R., "Rollover Testing on an Actual Highway," *SAE Int. J. Passeng. Cars - Mech. Syst.* 2, no. 1 (2009): 1237-1267, doi:10.4271/2009-01-1544.
4. Funk, J. and Luepke, P., "Trajectory Model of Occupants Ejected in Rollover Crashes," SAE Technical Paper 2007-01-0742, 2007, doi:10.4271/2007-01-0742.
5. Funk, J., Beauchamp, G., Rose, N., Fenton, S. et al., "Occupant Ejection Trajectories in Rollover Crashes: Full-Scale Testing and Real World Cases," *SAE Int. J. Passeng. Cars - Mech. Syst.* 1, no. 1 (2009): 43-54, doi:10.4271/2008-01-0166.
6. Perl, T., Bready, J., Nordhagen, R., and Warner, M., "Glass Debris in Rollover Accidents," *SAE Int. J. Passeng. Cars - Mech. Syst.* 1, no. 1 (2009): 55-59, doi:10.4271/2008-01-0167.
7. Przybyla, J., Jupe, J., Rush, T., and Keller, R., "Glass Debris Field Longevity for Rollover Accident Reconstruction," SAE Technical Paper 2015-01-1427, 2015, doi:10.4271/2015-01-1427.
8. Tandy, D.F., Pascarella, R.J., Colborn, J., and Hoover, T.D., "An Accident Reconstructionists Primer on Tire and Wheel Rim Marks at Crash Scenes," *Collision: The International Compendium of Crash Research*, ISSN 1934-8681.
9. Rose, N.A., Neale, W.T.C., Fenton, S.J., Hessel, D. et al., "A Method to Quantify Vehicle Dynamics and Deformation for Vehicle Rollover Tests Using Camera-Matching Video Analysis," *SAE Int. J. Passeng. Cars - Mech. Sys.* 1, no. 1 (2008): 301-317, doi:4271/2008-01-0350.
10. Orlowski, K., Moffatt, E., Bundorf, R., and Holcomb, M., "Reconstruction of Rollover Collisions," SAE Technical Paper 890857, 1989, doi:10.4271/890857.
11. Bready, J., May, A., and Allsop, D., "Physical Evidence Analysis and Roll Velocity Effects in Rollover Accident Reconstruction," SAE Technical Paper 2001-01-1284, 2001, doi:10.4271/2001-01-1284.
12. Turriff, D., King, D., and Bertoch, J., "On the Directionality of Rollover Damage and Abrasions," SAE Technical Paper 2015-01-1421, 2015, doi:10.4271/2015-01-1421.
13. Bandstra, R.A., "Damage in Opened and Closed Doors and Surrounding Structures in Rollovers," *Accident Investigation Quarterly*, no. 6, Spring 1995.
14. Raphael, E., Piziali, R., Le, H., Hinger, J. et al., "Physical Evidence Associated with Seatbelt Entanglement during a Collision," SAE Technical Paper 2007-01-1501, 2007, doi:10.4271/2007-01-1501.

15. Baker, K.S., "Chapter 9: Photogrammetry for Collision Analysis," in *Traffic Collision Investigation*, 10th edn., (Evanston, IL, Northwestern University Center for Public Safety, 2002), ISBN 0-912642-09-2.

16. Tumbas, N.S., "Photogrammetry and Accident Reconstruction: Experimental Results," SAE Technical Paper 940925, 1994, doi:10.4271/940925.

17. Pepe, M.D., "Accuracy of Three-Dimensional Photogrammetry as Established by Controlled Field Tests," SAE Technical Paper 930662, 1993, doi:10.4271/930662.

18. Niederer, P., Birchler, B., Mesqui, F., and Lehareinger, Y., "Computer Assisted Single-View Photogrammetry for Accident Scene Documentation," SAE Technical Paper 850067, 1985, doi:10.4271/850067.

19. Husher, S.E., Varat, M.S., and Kerkhoff, J.F., "Survey of Photogrammetric Methodologies for Accident Reconstruction," *Proceedings of the Canadian Multidisciplinary Road Safety Conference VII*, Vancouver, BC, Canada, June 1991.

20. Neale, W.T.C., Marr, J.P., and Hessel, D.R., "Video Projection Mapping Photogrammetry through Video Tracking," SAE Technical Paper 2013-01-0788, 2013, doi:10.4271/2013-01-0788.

21. Neale, W.T.C., Hessel, D., and Terpstra, T., "Photogrammetric Measurement Error Associated with Lens Distortion," SAE Technical Paper 2011-01-0286, 2011, doi:10.4271/2011-01-0286.

22. Massa, D.J., "Using Computer Reverse Projection Photogrammetry to Analyze an Animation," SAE Technical Paper 1999-01-0093, 1999, doi:10.4271/1999-01-0093.

23. Fenton, S., Neale, W., Rose, N.A., and Hughes, C., "Determining Crash Data Using Camera-Matching Photogrammetric Technique," SAE Technical Paper 2001-01-3313, 2001, doi:10.4271/2001-01-3313.

24. Chou, C., McCoy, R., Fenton, S., Neale, W. et al., "Image Analysis of Rollover Crash Test Using Photogrammetry," SAE Technical Paper 2006-01-0723, 2006, doi:10.4271/2006-01-0723.

25. Coleman, C., Tandy, D., Colborn, J., and Ault, N., "Applying Camera Matching Methods to Laser Scanned Three Dimensional Scene Data with Comparisons to Other Methods," SAE Technical Paper 2015-01-1416, 2015, doi:10.4271/2015-01-1416.

26. Breen, K.C. and Anderson, C.E., "The Application of Photogrammetry to Accident Reconstruction," SAE Technical Paper 861422, 1986, doi:10.4271/861422.

27. Woolley, R.L., White, K.A., Asay, A.F., and Bready, J.E., "Determination of Vehicle Crush from Two Photographs and the Use of 3D Displacement Vectors in Accident Reconstruction," SAE Technical Paper 910118, 1991, doi:10.4271/910118.

28. Main, B.W., "A New Application of Camera Reverse Projection in Reconstructing Old Accidents," SAE Technical Paper 950357, 1995, doi:10.4271/950357.

29. Campbell, A.T., "Adapting Three-Dimensional Animation Software for Photogrammetry Calculations," SAE Technical Paper 930904, 1993, doi:10.4271/930904.

30. Smith, G.C. and Allsop, D.L., "A Case Comparison of Single-Image Photogrammetry Methods," SAE Technical Paper 890737, 1989, doi:10.4271/890737.

31. Fenton, S. and Kerr, R., "Accident Scene Diagramming Using New Photogrammetric Technique," SAE Technical Paper 970944, 1997, doi:10.4271/970944.

32. Fenton, S., Johnson, W., LaRocque, J., Rose, N. et al., "Using Digital Photogrammetry to Determine Vehicle Crush and Equivalent Barrier Speed," SAE Technical Paper 1999-01-0439, 1999, doi:10.4271/1999-01-0439.

33. Switzer, D.A. and Candrlic, T.M., "Factors Affecting the Accuracy of Non-Metric Analytical 3-D Photogrammetry, Using PhotoModeler," SAE Technical Paper 1999-01-0451, 1999, doi:10.4271/1999-01-0451.

34. Peck, L. and Cheng, M., "The Accuracy of an Optimized, Practical Close-Range Photogrammetry Method for Vehicular Modeling," *SAE Int. J. Trans. Safety* 4, no. 2 (2016): 245-266, doi:10.4271/2016-01-1462.

CHAPTER 3

Analysis Methods: Rollover Phase

Average Deceleration Rates

Reconstructionists have often employed the assumption that a rolling vehicle decelerates at a constant rate. While, in reality, this is not the case, this assumption can yield an accurate calculation of the vehicle speed when the roll phase begins. Average deceleration rates (drag factors) for the roll phase can be calculated from rollover crash tests using on the following equation:

$$f_{roll} = \frac{1}{2}\frac{v_{roll}^2}{gd_{roll}} \tag{3.1}$$

In Equation (3.1)

v_{roll} is the over-the-ground speed of the vehicle at the beginning of the roll phase
d_{roll} is the distance the vehicle rolls from the beginning of the roll phase to rest
g is the gravitation acceleration
f_{roll} is the drag factor or deceleration rate in gravitational units

A 1972 study by Hight [1] recommended using a range of 0.4 to 0.65 g for the deceleration rate of a rolling vehicle. Hight studied 139 rollover collisions, 90 of which involved a single vehicle. He reported that "analysis was made of the average deceleration of vehicles during the rolling phase of the collision. Various factors were included in establishing the rollover speed, such as: other road users' statements, highway geometry, braking and centrifugal skid marks, critical cornering speeds, etc. After establishing the cruise speed, the estimated rollover speed was then obtained for about 70% of the cases. In the other cases, there was insufficient physical evidence available to estimate the speed with a reasonable degree of certainty. Sixty percent of the vehicles that rolled on approximately level ground decelerated between the range of 0.40-0.65g."

In 1985, Orlowski et al. [2] reported eight dolly rollover tests conducted with 1983 Chevrolet Malibus at speeds of approximately 32 mph (51.5 km/h). These Malibus were front-engine, rear-wheel-drive cars that weighed approximately 3,200 lb. Four of the vehicles had standard, production roofs and four had modified roofs with roll cages. In these tests, the vehicles were launched laterally onto a flat and dry asphaltic concrete surface with their passenger (right) sides leading and with an initial inclination of 23°. Unrestrained 50th percentile Hybrid III dummies were situated in the driver's and passenger's side front seats. In these tests, the vehicles with standard roofs rolled between 2½ and 3½ revolutions over distances between 66 and 89 ft (20.1 to 27.4 m). The vehicles with roll cages rolled between 2 and 3½ revolutions over distances between 65 and 74 ft (19.8 and 22.6 m). From the launch point to the point of rest, the average deceleration rate for these tests was 0.43 g. In 1989, Orlowski [3] presented an analysis of 41 additional dolly rollovers crash tests and reported that the average deceleration rate for these tests was 0.42 g, with a range of 0.36 to 0.61 g.

In a 2011 study, Arndt [4] reevaluated Hight's deceleration data. He noted that it contained 102 data points and that the full range of deceleration rates was 0.04 to 1.20 g. This range included the rollovers that occurred on a downhill surface and rollovers with vertical drops (hence the wide range of deceleration rates). Arndt reported that the full range of deceleration rates for rollovers on flat ground in Hight's study was 0.21 to 0.83 g. Arndt suggested discontinuing use of the Hight study as a basis for rolling vehicle deceleration rates, observing that while "the Hight…study made a significant historical contribution to the science of rollover analysis…no experimental program has documented rollover deceleration at the 0.65g level without unusual circumstances…the Hight study was based on reconstruction techniques without experimental input from the 1960's."

Arndt also noted differences between dolly rollover tests and steering-induced rollovers. For example, for analysis of steering-induced rollovers, the roll phase is considered to begin where the tire marks or furrows of the trip phase end. The first impact with the ground during the roll phase is typically with the roof. During a typical dolly rollover test, the vehicle exits the dolly from an elevated position, with an initial roll angle and a roll velocity introduced through the cart deceleration and a lip on the cart. The process of the vehicle exiting the dolly initiates the trip phase. However, when the vehicle lands, it typically lands first with its leading side wheels. This impact between the leading side wheels and the ground decelerates the vehicle further and introduces additional roll velocity. Thus, this first impact could be considered a continuation of the tripping of the vehicle. When this wheel impact comes to an end, the vehicle becomes airborne and the roll phase would then have begun in earnest. However, most of the deceleration rates that have been reported for dolly rollovers have used either the vehicle speed at launch from the dolly and the corresponding roll distance from the point launch from the dolly or the vehicle speed when the wheels first touch down and the corresponding roll distance from this point. Arndt makes the case that to calculate a deceleration rate from a dolly rollover that is consistent with the roll phase for a naturally occurring rollover, one would need to use the speed and roll distance referenced to the end of the leading side tire impact with the ground. Frequently, neither of these parameters is documented for dolly rollover tests. The counterargument is that wheel impacts similar to what a vehicle experiences when it exits a dolly occur during the roll phase of naturally occurring rollovers, and so, there is unlikely to be much difference. This seems borne out by the deceleration rates, Arndt actually reports for naturally occurring rollovers when these are compared to those reported for dolly rollovers.

Luepke and Asay [5] compared the dynamics of dolly rollovers to steering-induced rollover tests. They acknowledged that "no rollover event in the field is likely to be initiated by the sudden deceleration of a large dolly cart holding a subject vehicle at a roll angle of 23 degrees." However, they also pointed out that "once a test vehicle leaves the dolly, the only parameters influencing or controlling the dynamics of the vehicle are gravity, surface environment, and the laws of physics…The question that remains is then, what influence does rollover initiation on a dolly have on the subsequent rollover dynamics of a vehicle?" In attempting to answer this question, Luepke and Asay compared eight dolly rollover tests [6, 7, 8] and seven steer-induced rollover tests [9] – all involving a sport utility vehicle or multipurpose vehicle and all involving rollovers that occurred, at least partially, on dirt.

Luepke and Asay observed that "it quickly became apparent that despite two completely different methods of rollover initiation, the overall results of each rollover had remarkable correlation to the other rollover events. Relationships between the initial velocity and the…total rolls and total distance existed, and can be described as near-linear…in general, a higher velocity resulted in more rolls and a greater rollover distance, independent of how the rollover was initiated…no evidence exists to suggest that the dolly rollover method introduces an artificially higher number of rolls for a similar total rollover distance…The data not only clearly supports a robust relationship between the dolly rollover method and natural rollovers, but it provides important insight into the remarkably homogeneous effect that the dirt rollover surface has upon vehicle rollover dynamics." Luepke and Asay concluded: "Dirt dolly rollovers, when compared to steer-induced off road rollovers, produce substantially similar average distances per complete roll. The roll rate histories of dirt dolly rollovers are remarkably similar to off-road steer-induced rollovers in shape, duration, magnitude of peak, time-to-peak, and time-to-zero."

Arndt [4] reexamined the results of numerous published rollover tests – 81 dolly rollover tests, 24 naturally occurring rollovers (18 of which were steering-induced), and 102 reconstructed rollovers. He identified several instances where the original reported results for the speed at the beginning of the roll had been overestimated – particularly in Asay's 2010 study. He corrected these results and recalculated the average deceleration rates for the tests. For the dolly rollovers he compiled, Arndt found that the drag factors ranged between 0.38 and 0.50 g with the upper and lower 15% statistically trimmed. The mean value was 0.44 g with a standard deviation of 0.064 g. For the naturally occurring rollovers considered by Arndt, he found that the drag factors ranged between 0.39 and 0.50 g with the upper and lower 15% statistically trimmed. The average value was 0.44 g with a standard deviation of 0.063. Thus, comparison of deceleration rates from dolly rollovers and naturally occurring rollovers supports the contentions of Luepke and Asay.

Once the roll phase distance has been determined, these average deceleration rates can be applied within the following equation to calculate the vehicle's speed when the roll phase began:

$$v_{roll} = \sqrt{2gf_{roll}d_{roll}} \tag{3.2}$$

Determining Likely Number of Rolls from Roll Distance

After calculating the speed of a vehicle at the beginning of the roll phase, the likely number of rolls given the roll distance can then be considered. Figure 3.1 is a graph plotting the roll distance versus the number of rolls for naturally occurring rollovers reported in Larson [10], Wilson [11, 12], Anderson [13], Asay [9, 14], Stevens [15], Asay [16], and Larson [17]. This data includes vehicles of all types (passenger cars, pickups, minivans, and sport utility vehicles) on various roll surfaces. There are 21 total data points. A linear relationship has been fit to this data with the result that one roll occurs approximately every 30 ft (9.1 m) of roll distance. This is, of course, only an average. Each of these tests exhibits variability in the roll distance per roll over the course of the test and, even on an average distance per roll basis, there is considerable variability from test to test. If one chooses to use Figure 3.1 to estimate the likely number of rolls for a particular roll distance before beginning to examine the physical evidence, one should keep in mind that the number of rolls that results from the equation on the graph of Figure 3.1 yields a value that is roughly ±2 rolls. An SI unit version of Figure 3.1 is included in the appendix.

Figure 3-2 is similar to Figure 3.1 but it only contains the data for sport utility vehicles. There are 12 data points on this graph. Fitting a linear relationship to this data results in approximately the same equation as the one in Figure 3.1. However, there is less variability. Within the data for sport utility vehicles only, one could potentially predict the actual number of rolls, ±1 roll, just given the roll distance. Ultimately, of course, the physical evidence should drive the reconstruction of the roll phase, but it may be helpful to have an idea of the approximate number of rolls going into the evidence analysis. An SI unit version of Figure 3-2 is included in the appendix.

Several authors have reported data related to the number of rolls that occur for a given distance based on their own reconstructions. In general, the reconstructed rollovers had fewer rolls for a given distance than the full-scale tests compiled in Figure 3.1. This could be due, in part, to the scratch evidence on the vehicles representing only a minimum number of rolls. Also, lower-speed dolly rollovers in early testing suggested average distances of 50 ft per roll.

FIGURE 3.1 Number of rolls versus roll distance (naturally occurring rollovers, all vehicle types and surfaces).

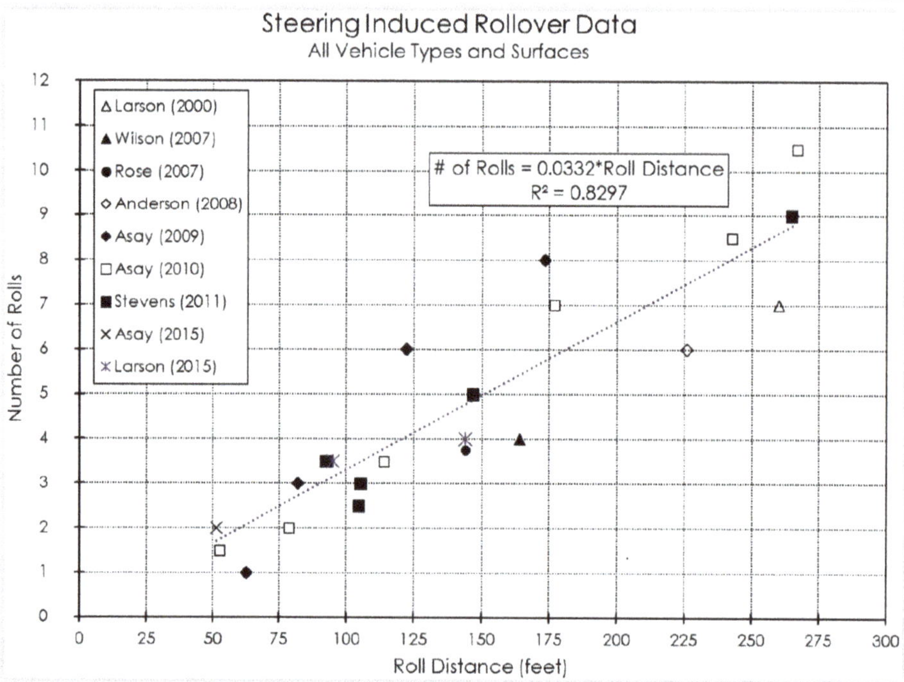

FIGURE 3-2 # of Rolls versus roll distance (naturally occurring rollovers, SUVs, all surfaces).

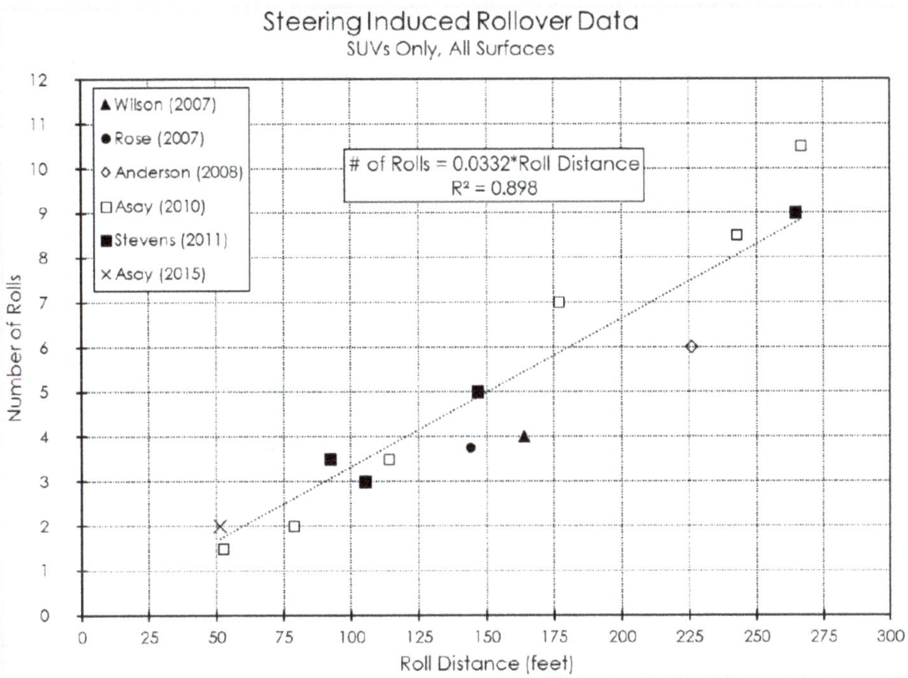

Jones and Wilson [18] reported data for 38 reconstructed rollovers involving sport utility vehicles, light trucks, and minivans. There is considerable scatter in the Jones and Wilson data. For instance, in this data, three complete revolutions of the vehicle could occur over distances between 100 and 225 ft and four complete revolutions could occur over distances between 150 and 400 ft. The Jones and Wilson data shows fewer rolls for any given distance than the naturally occurring rollover crash test data reported in Figures 3-2 and 3.3.

Altman et al. [19] presented data for 24 reconstructed rollovers involving passenger cars, sport utility vehicles, and pickups. The data shows less scatter than the Jones and Wilson data, but the level of scatter is consistent with the level of scatter in the data of Figures 3.1 and 3-2. The Altman data also shows fewer rolls for any given distance than the naturally occurring rollover crash test data reported in Figures 3.1 and 3-2.

Herrera and Najera [20] reported data for 90 reconstructed rollovers involving passenger cars, sport utility vehicles, minivans, and pickups. As with the other studies, there is considerable scatter in the Herrera and Najera data. For instance, in this data, three complete revolutions of the vehicle could occur over distances between 80 and 160 ft, and four complete revolutions could occur over distances between 110 and 210 ft. The Herrera and Najera data shows fewer rolls for any given distance than the naturally occurring rollover crash test data reported in Figures 3-2 and 3.3.

In addition to these studies, several authors have made the observation from the rollover data that there is a limit to the number of rolls that a vehicle could possibly achieve in a given distance (from trip to rest). The maximum number of rolls for a typical rollover can be estimated by taking the roll distance and dividing it by the rolling circumference of the vehicle (including the tires). This estimate of the maximum number of rolls would be based on the assumption of minimal translation, with the roll motion of the vehicle accounting for nearly all of the vehicle motion down the roll path.

FIGURE 3.3 Video Images from Case #3 from Rose and Beauchamp [21].

Since nearly all rollovers typically have some degree of translational velocity (or sliding) rather than just pure rotation, the division of the rollover distance by the vehicle circumference would provide a maximum estimate to the number of rolls.

Typical Rollover Characteristics

The data in the previous section enables a reconstructionist to statistically estimate the number of rolls a vehicle likely completed for a known roll distance. The physical evidence both on the ground and the vehicle drive the ultimate determination as to how many times a particular vehicle actually rolled. In determining a plausible match with the evidence, it is helpful to understand how a vehicle's roll velocity typically varies through the roll phase. This allows the analyst to generate a reconstruction that accounts for the physical evidence and also makes physical sense.

Rose and Beauchamp [21] analyzed the dynamics of 12 high-speed, real-world rollover crashes that were captured on video and used this analysis to explore the characteristics of typical roll velocity curves. For instance, consider the dynamics of Case #3 from Rose and Beauchamp, a high-speed, multiple-roll, soil-tripped rollover crash involving a GMC Yukon Denali. Figure 3.3 contains frames from the video of this crash. This vehicle rolled for approximately 144 ft (43.9 m) and had an over-the-ground speed at the beginning of the roll of 48 mph (77.4 kph) - an average deceleration rate of 0.53 g.

Figure 3.4 depicts the roll velocity curve for this crash, plotted with the progression of the 3-¾ rolls on the horizontal axis. After completing ¼ roll, the roll velocity of the vehicle was around 200°/s. By the time the vehicle completed its first roll, the roll velocity had increased to around 450 degrees per second. Roll velocities exceeding 400°/s were then maintained nearly through the third roll. From that point forward, the roll velocity generally decreased until the vehicle came to rest.

In analyzing the roll velocity of the vehicles in their study, Rose and Beauchamp counted the number of video frames over which the vehicles traversed each 90° interval of roll. Using a known frame rate, the number of frames for each 90° interval could be converted to time and the average roll velocity for each interval could be calculated. As this study noted, sampling at 90° intervals can result in significant smoothing of the

FIGURE 3.4 Roll Velocities from Case #3 from Rose and Beauchamp [21].

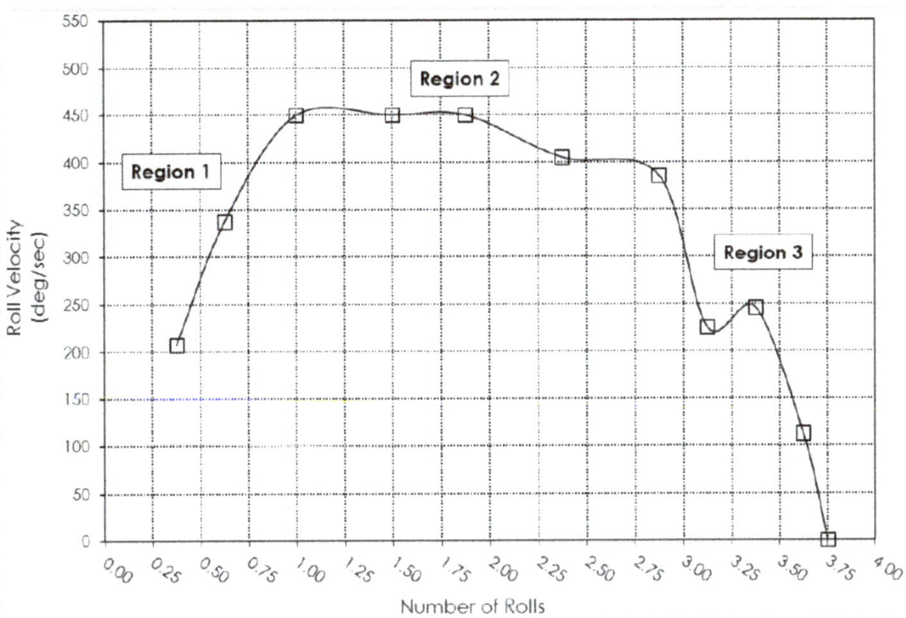

actual roll rate time histories. Rose and Beauchamp used crash test data and simulation to quantify the effects of this sampling interval and concluded that substantial changes in roll velocity that occurred quickly in the actual roll velocity data occurred much more slowly in a roll velocity curve generated with 90° or 180° roll angle intervals. In addition to this smoothing, the curves generated with these sampling intervals did a relatively poor job of capturing the buildup of roll velocity that occurred during the first ¼ roll. Rose and Beauchamp ultimately concluded, though, that the roll velocity time history generated with the 90° sampling provided a reasonable approximation of the actual roll velocity curve for roll rates under 450°/s. Generally, the actual instantaneous roll velocity curve was contained within a roll velocity envelope lying within ±50°/s of the average roll velocity curve from the video analysis.

The roll velocity history for the crash of Figures 3.3 and 3.4 is similar to that for other high-speed, multiple-roll crashes presented in Rose and Beauchamp's article. The roll velocities in a number of the cases reached a moderate level after the vehicle completed about ¼ roll (the beginning of the roll phase), then built up to a high roll velocity level. Often, these high roll velocities were maintained for some period of time before the roll velocity began to diminish prior to the vehicle coming to rest. The roll velocity history of Figure 3.4 can conceptually be split into the following three regions:

- **Region 1** – In this region, the vehicle's roll velocity builds up from its level coming out of the trip phase to a peak or near-peak level.
- **Region 2** – In this region, the roll velocity reaches a plateau, with high roll velocities generally maintained and the vehicle experiencing only small increases or decreases in roll velocity.
- **Region 3** – In this region, the roll velocity steadily diminishes until the vehicle comes to rest.

Rose and Beauchamp summarized data from the crashes they evaluated and their data is included in Table 3.1. Not all of the roll velocity histories for the rollovers analyzed by Rose and Beauchamp exhibited the plateauing of the roll velocity (Region 2).

TABLE 3.1 Rollover data reported by Rose and Beauchamp [21]

Case #	Vehicle type	Trip type	# of rolls	Initial roll rate (°/s)	Peak roll rate (°/s)	Average roll rate (°/s)
1	Pickup truck	Off road furrow	2	245	386	247
2	Large SUV	On road	1.75	270	415	338
3	Large SUV	Off road furrow	3.75	208	450	302
4	Rally car	Impact	4	540	573	272
5	Road race	Off road furrow	2.25	338	338	173
6	Rally car	Impact	3	169	476	261
7	Rally car	Ramping	5	169	491	310
8	Rally car	Drop off road	4	225	386	174
9	Rally car	Impact	1.5	270	300	190
10	Rally car	Impact	1.5	225	338	144
11	Rally car	Dirt road furrow	7	150	675	395
12	Rally car	Off road furrow	4	–	–	–

Studies by other authors have also revealed that Region 2 does not always occur and frequently the roll velocity history consists only of Regions 1 and 3. Funk [22] developed a theoretical model for the roll phase that predicted the roll velocity history would consist of only Regions 1 and 3. This model agreed well with the data from a number of rollover crash tests, including some of Stevens' [15] and Asay's [9] steering-induced tests. Funk observed, however, that several of the rollovers he analyzed did exhibit the plateauing region. He stated, "The physical interpretation of a plateau rather than a triangular peak in the roll rate data is that the vehicle transitioned from sliding to rolling upon landing from an airborne phase." Suffice it to say at this point that when reconstructing the roll phase, the physical evidence should drive the reconstruction. Region 2 of the roll velocity curve may or may not be present.

Mathematical Analysis of the Orientation of Scratches on the Vehicle

When the body panels of a vehicle contact the ground, scratch marks are often created on these panels. In combination with other evidence, these scratch marks can be used to determine the yaw orientation of the vehicle at the point during the roll when the scratches were created. Also, on body panels where there are multiple families of scratches with different orientations, overlapping scratches can demonstrate how many times that part of the vehicle was in contact with the ground during the rollover [3]. This is an indication of the minimum number of rolls experienced by the vehicle during the crash – minimum because the vehicle continues to roll while it is airborne [9, 23], and also, soft soil or grass surfaces may not deposit scratch marks on the body panels [3].

As mentioned in Chapter 2, Bready [23] raised a potential problem with using the orientation of scratch marks to determine the yaw orientation of the vehicle, noting that "evidence from the scene and vehicle...can sometimes appear inconsistent, because they suggest a substantially different vehicle orientation at a particular contact point." Bready suggests that this inconsistency is due to the incorrect assumption that the scratch direction can simply be aligned to the over-the-ground velocity direction of the vehicle to determine the vehicle orientation. "If a vehicle were simply sliding on the ground without any rolling motion, then the resulting scrapes would be parallel to, and opposite in direction of, the vehicle horizontal velocity... If a vehicle was only rotating, and not translating over the ground, then the resulting scrapes would be parallel to, and opposite in direction of, the roll motion of the vehicle, and its associated peripheral velocity. However, there are generally both roll and translational motions in a rollover accident. The contribution of these two components determines the actual velocity of the vehicle contact point, relative to the ground, and will determine the angle and direction of the scrape marks left on the vehicle...the angle and direction of the scrapes and scratches are a record of the relative motion of the contacting vehicle body surfaces with respect to the ground. In order to correctly orient the vehicle using the documented scrape information, the analyst may need to determine the contribution of the roll and translational velocities on scrape angle during ground contacts. When the effects of roll velocity are used in conjunction with traditional methods, an apparent inconsistency between evidence may be resolved and the analyst should be better able to determine the actual vehicle yaw angle in a rollover accident."

Bready presented the following equation that yields the scratch orientation, relative to the roll axis of the vehicle, for a particular combination of vehicle slip angle (α), translational velocity at the center of mass (v_{trans}), and perimeter velocity (v_{roll}). The perimeter velocity is calculated by multiplying the roll velocity (ω_{roll}) by the distance from the center of mass to the vehicle perimeter (r_{roll}). In the example presented in Bready's paper, he estimated r_{roll} by calculating the radius of a circle with a perimeter equal to twice the vehicle height plus twice its width. In practice, the reconstructionist would first calculate the vehicle's over-the-ground speed and roll velocity over the course of the roll phase. The slip angle of the vehicle throughout the roll would also be estimated. Then, the scratch angle associated with particular points along the roll path could be calculated and compared to the actual scratch patterns on the vehicle. The vehicle slip angles would then be iterated until the orientation of the scratch patterns on the vehicle agree with the calculated scratch directions:

$$\tan\theta = \frac{\sin\alpha - \left(\frac{v_{roll}}{v_{trans}}\right)}{\cos\alpha} = \frac{\sin\alpha - \left(\frac{\omega_{roll}\cdot r_{roll}}{v_{trans}}\right)}{\cos\alpha} \tag{3.3}$$

Reconstruction Methodology - Roll Phase (Constant Deceleration)

Following is a step-by-step procedure for reconstructing the roll phase. This procedure is summarized graphically in Figure 3.5:

1. **Map the scene and vehicle evidence.** For the scene, this typically involves visiting the site and/or conducting photogrammetric analysis. Then, an evidence diagram is often created in a CAD program. This diagram can include the geometry of the on-road and off-road surfaces and also the physical evidence deposited on the ground during the rollover. This diagram often combines information and measurements obtained from police measurements, on-site documentation, and photogrammetric analysis. For the vehicle, this typically involves inspecting the vehicle, or at least analyzing photographs of the vehicle damage. Also, vehicle specifications are typically obtained and a computer model of the precrash shape of the vehicle created to accompany the evidence diagram. Calculate the center of mass location and the moments of inertia for the vehicle. In addition, consider inspecting an exemplar vehicle to

assist with identifying key components or structure, especially for the instance that the accident vehicle is not available.

2. **Determine the roll distance.** This is often accomplished by using the evidence diagram to place one vehicle model at the point of rest and another on the end of the trip phase tire marks, oriented at the neutral stability angle. The distance between these vehicles can then be measured.
3. **Assign a deceleration rate for the roll phase.** Typically, this deceleration rate falls in the range of 0.38 to 0.50 g.
4. **Calculate the speed at the beginning of the roll phase** with Equation (3.4). The total duration of the roll phase can also be calculated with Equation (3.5):

$$v_{roll} = \sqrt{2gf_{roll}d_{roll}} \tag{3.4}$$

$$t_{roll} = \frac{2 \cdot d_{roll}}{v_{roll}} \tag{3.5}$$

5. **Determine vehicle positions and orientations for the roll phase.** These positions are established by matching the data on the evidence diagram with the damage on the vehicle. This evidence includes ground impact marks, glass deposits, and scratch marks.
6. **Calculate the velocity at each roll phase position** using Equation (3.6), beginning from the beginning of the roll phase and working towards the rest position. In Equation (3.6), the *i* position represents the upstream position for any particular interval and the *i* – 1 position represents the downstream position for the interval. Thus, Equation (3.6) starts with the speed of the vehicle at the beginning of the interval and yields the speed of the vehicle at the end of the interval based on the distance of the interval and the roll phase deceleration rate:

$$v_{i-1} = \sqrt{v_i^2 - 2gf_{roll}\Delta d_{i\to i-1}} \tag{3.6}$$

7. **Calculate the time between roll phase positions** using Equation (3.7):

$$\Delta t_{i\to i-1} = \frac{2\Delta d_{i\to i-1}}{v_i + v_{i-1}} \tag{3.7}$$

8. **Calculate the average roll velocity for each segment of the roll phase and construct a roll velocity curve.** Equation (3.8) can be used for this purpose. It uses the change in roll angle and the time for the interval to calculate the average roll velocity of the vehicle over the interval:

$$\omega_{avg,i-1\to i} = \frac{\theta_{i-1} - \theta_i}{\Delta t_{i-1\to i}} \tag{3.8}$$

This is an iterative process. The positions established with physical evidence essentially remain fixed, but there is likely to be some uncertainty in the roll orientations for these positions that affect the shape of the roll velocity curve (adjustment of ±10°). Additional roll positions may also need to be established based on the roll velocity curve. The goal is a reconstruction of the roll phase that is consistent with the physical evidence and that also yields a physically realistic roll velocity curve. While working towards a reasonable roll velocity curve, check

FIGURE 3.5 A step-by-step procedure for the roll phase.

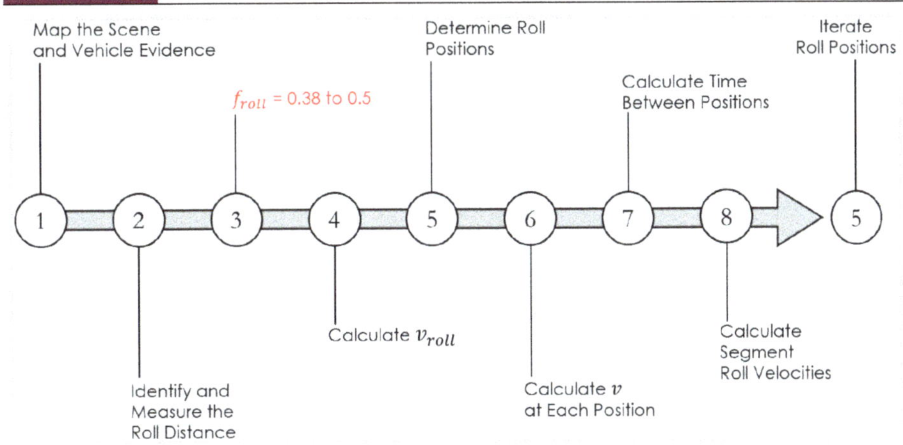

the scratch patterns on the vehicle for their consistency with the reconstruction. When combined with an analysis of the evidence on the ground, scratch pattern analysis could potentially reveal which portions of the roll the vehicle were in contact with the ground and which portions of it were airborne.

Case Study: Asay's Isuzu Rodeo Rollover Test (Test #2, 2010-01-0521)

This section demonstrates the methodology outlined in the previous section using the steering-induced rollover tests reported by Asay in 2010. This particular test utilized a 1991 Isuzu Rodeo. This test was conducted on the same remote, rural highway in the west desert of Utah on which nearly all of Asay's tests were conducted. In this test, the Isuzu was towed up to and released at a speed of 73.5 mph (118.3 kph). After release, the vehicle was steered with a sharp left steering input of approximately ¼ turn, causing the vehicle to travel to the left across the roadway and to yaw counterclockwise, developing a slip angle of sufficient magnitude that tire marks were deposited on the road surface. Then, 1 s later, the leftward steering input was followed by a severe steering input back to the right. The vehicle continued off the left side of the road into the dirt but reversed its yaw direction, developing a significant slip angle as it yawed in a clockwise manner. The vehicle deposited furrows in the dirt and then began rolling over. The vehicle rolled 7 times in 181 ft (55.3 m). Asay surveyed the physical evidence both on and off the roadway from this test. In this case, the physical evidence included tire marks, tire furrows, disturbed earth, wheel landings, glass spills, and debris from the vehicle. Based on Asay's survey data, the evidence diagram of Figure 3.6 was produced.

FIGURE 3.6 Evidence diagram created for Asay's Isuzu Rodeo test.

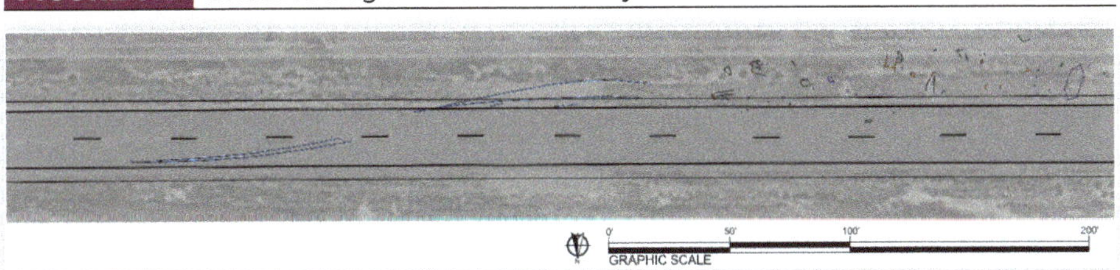

FIGURE 3.7 Roll positions determined from the scene evidence.

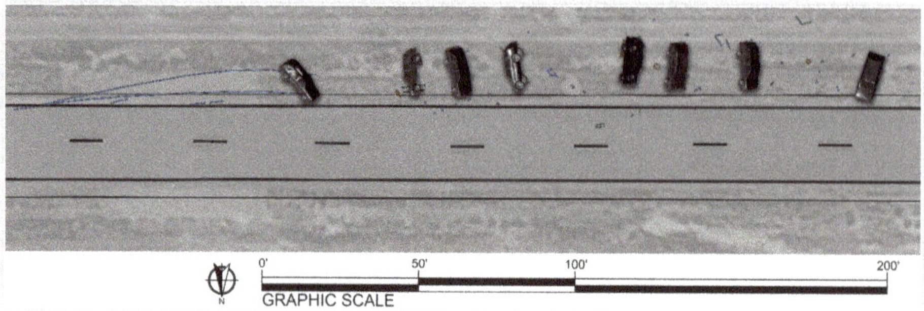

FIGURE 3.8 Roll phase measurements.

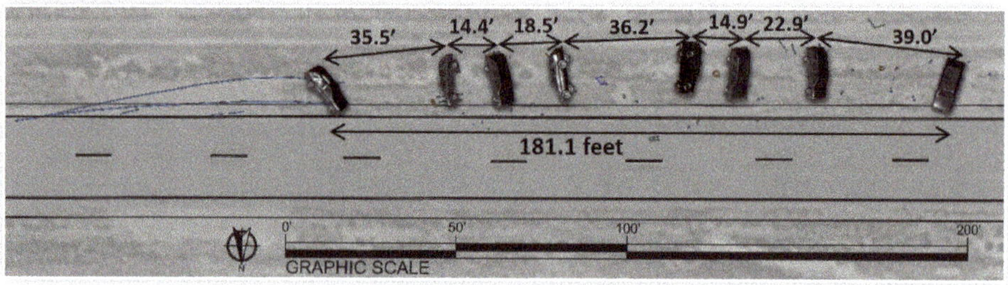

Using the physical evidence diagram of Figure 3.6 along with a computer model of the Isuzu, the eight vehicle positions depicted in Figure 3.7 were determined. The first of these positions occurs at or near the end of the trip phase (oriented at the neutral stability angle for the vehicle), and the last position is the rest position of the vehicle. The six positions between these are roll positions of the vehicle determined from physical evidence on the ground. Figure 3.8 depicts these same 8 vehicle positions, but in this figure, measurements between the positions are depicted. These measurements resulted in a total roll distance of 181 ft.

Using the measured roll distance of 181 ft (55.2 m) and a typical deceleration rate of 0.44 g for the roll phase, the speed of the vehicle at the beginning of the roll phase can be calculated as follows:

$$v_{roll} = \sqrt{2 \cdot 32.2 \cdot 0.44 \cdot 181.3} = 73.1\, fps = 48.9\,\text{mph}$$

$$t_{roll} = \frac{2 \cdot 181.3}{73.1} = 5.05\,\text{s}$$

Asay [9] reported a speed of the Rodeo at the beginning of the roll phase (trip) of 50.7 mph. Arndt [4] reevaluated Asay's analysis and reported a revised speed for the beginning of the roll phase of 44.4 mph. These authors have reevaluated the speed at the beginning of the roll phase, and our results generally agree with Arndt's. With a roll distance of 181 ft, a speed of 44.4 mph implies an average deceleration rate of 0.363 g, outside of the normal range of typical values. Thus, in this instance, using a typical deceleration rate of 0.44 g for the roll phase overestimates the actual roll speed by 4.5 mph. Rerunning the previous calculations with the actual deceleration rate of 0.363 results in the following:

$$v_{roll} = \sqrt{2 \cdot 32.2 \cdot 0.363 \cdot 181.3} = 44.4 \text{ mph}$$

$$t_{roll} = \frac{2 \cdot 181.3}{44.4 \cdot 1.467} = 5.56 \text{ s}$$

Based on the data from the roll velocity sensor on the vehicle, the actual duration of the roll phase in this case was just more than 5½ s, so using the actual deceleration rate also yields a more accurate prediction of the roll duration. Using a typical average deceleration rate for the roll phase of 0.44 g results in the roll duration being underestimated by about ½ s.

The linear fit reported in Figure 3.1 predicts that for a roll distance of 181 ft, a vehicle would have rolled 6 times - or between 5 and 7 times, considering the scatter in the data. This raises the question of whether other evidence and analysis would enable the reconstructionist to determine that there were actually 7 rolls in this case, as opposed to 5 or 6. Review of Asay's documentation of the Isuzu after the test reveals that he identified 6 scratch families. The maximum number of scratch families on a single side of the vehicle was 5. This is consistent with the idea that scratch patterns give an indication of the minimum number of rolls, not necessarily the actual number. The video from this test shows that the vehicle did, in fact, experience a complete roll while it was airborne, and so, there would be no scratches on the vehicle or marks on the ground evidencing this portion of the roll.

The 8 positions of Figures 3.7 and 3.8 define 7 segments during the roll phase to which Equations (3.6) through (3.8) can be applied to determine vehicle speeds, segment times, and roll velocities. These calculations are carried out and summarized below in Table 3.2 using a deceleration rate of 0.363 g. For each segment, this table lists the segment distance, in both feet and meters, and the roll angle at the beginning and end of each segment in degrees. Calculations are then carried out, again assuming a constant deceleration rate of 0.363 g, to yield the vehicle speed, in mph and kph; the segment duration,

TABLE 3.2 Roll phase calculations for Asay's Isuzu Rodeo test

	Segment distance		Roll angle (°)	Vehicle speed (mph/kph)		Segment duration (s)	Segment roll velocity (°/s)
End of trip			48	44.4	71.5		
1	35.4 ft	10.8 m				0.57	388
			270	39.9	64.2		
2	14.4 ft	4.4 m				0.25	495
			395	37.8	60.8		
3	18.5 ft	5.6 m				0.35	578
			595	35.1	56.5		
4	36.2 ft	11.0 m				0.77	791
			1205	28.9	46.5		
5	14.9 ft	4.5 m				0.37	716
			1470	26.0	41.8		
6	22.9 ft	7.0 m				0.67	559
			1845	20.6	33.2		
7	39.0 ft	11.9 m				2.57	263
Rest			2520	0			
Totals	181.3 ft	55.3 m				5.56	

CHAPTER 3

FIGURE 3.9 Roll velocity curves calculated assuming 6 and 7 rolls compared to the actual roll velocities from the sensor data.

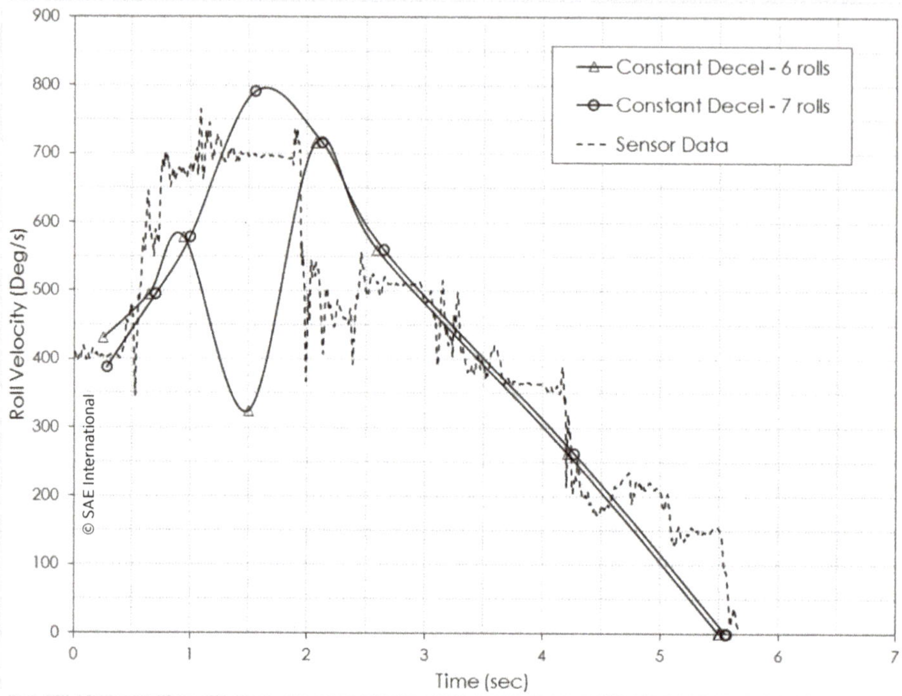

in seconds; and the average roll velocity for the segment, in degrees per second. Figure 3.7 and a partially filled out version of Table 3.2 are also included in the Chapter 3 exercises where the reader can practice these calculations.

The roll angles listed in Table 3.2 were determined from the physical evidence but also benefited from review of the video from this test. Thus, these roll angles reflect 7 complete rolls by the vehicle. The question remains as to whether additional evidence and analysis would enable the reconstructionist to identify the presence of 7 rolls in this case (without the benefit of video) even though one of those rolls occurred while the vehicle was airborne. To answer this question, consider the graph of Figure 3.9. This figure is a roll velocity graph that contains three curves – the sensor data that gives the measured roll velocities, the reconstructed roll velocity curve assuming 6 rolls, and the reconstructed roll velocity curve assuming 7 rolls. The airborne roll occurred during the 4th segment, and so the difference between the reconstructed curves is reflected in the roll velocity calculated for this segment. The reconstructed curve that assumed 6 rolls exhibits a precipitous drop in roll velocity below its value for the 3rd and 5th segments. The shape of this reconstructed curve does not reflect the characteristics of typical roll velocity curves discussed earlier in this chapter. The reconstructionist could reasonably reject this curve. The reconstructed roll velocity curve that assumed 7 rolls does have the appearance of a typical roll curve. Thus, in this case, a reconstructionist could detect the airborne roll even though there would be no physical evidence associated with it.

As an additional check, Asay reported the roll circumference of the Rodeo was 22.2 ft (6.77 m). Dividing the roll distance of 181 ft (55.2 m) by 22.2 ft yields 8.2 for the maximum number of rolls that the Rodeo could have experienced over that distance. When considering the evidence of translation along with the airborne phase, the actual number of rolls would have to have been less than 8 rolls.

References

1. Hight, P., Siegel, A., and Nahum, A., "Injury Mechanisms in Rollover Collisions," SAE Technical Paper 720966, 1972, doi:10.4271/720966.
2. Orlowski, K., Bundorf, R., and Moffatt, E., "Rollover Crash Tests-The Influence of Roof Strength on Injury Mechanics," SAE Technical Paper 851734, 1985, doi:10.4271/851734.
3. Orlowski, K., Moffatt, E., Bundorf, R., and Holcomb, M., "Reconstruction of Rollover Collisions," SAE Technical Paper 890857, 1989, doi:10.4271/890857.
4. Arndt, M., "Drag Factors from Rollover Crash Testing for Crash Reconstructions," *Proceedings of the ASME 2011 International Mechanical Engineering Congress & Exposition IMECE2011*, Denver, CO, November 11-17, 2011, IMECE2011-65537.
5. Luepke, P. and Asay, A., "Comparing Dolly Rollover Testing to Steer-Induced Rollover Events for an Enhanced Understanding of Off-Road Rollover Dynamics," *SAE Int. J. Passeng. Cars - Mech. Syst.* 4, no. 1 (2011): 848-869, doi:10.4271/2011-01-1112.
6. Luepke, P., Carhart, M., Croteau, J., Morrison, R. et al., "An Evaluation of Laminated Side Window Glass Performance During Rollover," SAE Technical Paper 2007-01-0367, 2007, doi:10.4271/2007-01-0367.
7. Luepke, P., Carter, J., Henry, K., Germane, G. et al., "Rollover Crash Tests on Dirt: An Examination of Rollover Dynamics," *SAE Int. J. Passeng. Cars - Mech. Syst.* 1, no. 1 (2009): 18-30, doi:10.4271/2008-01-0156.
8. Croteau, J., Zolock, J., Larson, R., Bare, C. et al., "Dynamic Response of Vehicle Roof Structure and ATD Neck Loading during Dolly Rollover Tests," *SAE Int. J. Passeng. Cars - Mech. Syst.* 3, no. 1 (2010): 407-449, doi:10.4271/2010-01-0515.
9. Asay, A. and Woolley, R., "Rollover Testing of Sport Utility Vehicles (SUVs) on an Actual Highway," SAE Technical Paper 2010-01-0521, 2010, doi:10.4271/2010-01-0521.
10. Larson, R., Smith, J., Werner, S., and Fowler, G., "Vehicle Rollover Testing, Methodologies in Recreating Rollover Collisions," SAE Technical Paper 2000-01-1641, 2000, doi:10.4271/2000-01-1641.
11. Wilson, L.A., Gilbert, M., and Godrick, D., "Reconstruction and Analysis of Steering-Induced, On-Road, Untripped SUV Rollover Tests (Part 1)," in *Collision—The International Compendium for Crash Research* 2, no. 1, Spring 2007, ISBN 1934-8681.
12. Wilson, L.A., Gilbert, M., and Godrick, D., "Reconstruction and Analysis of Steering-Induced, On-Road, Untripped SUV Rollover Tests (Part 2)," in *Collision—The International Compendium for Crash Research* 2, no. 2, Fall 2007, ISBN 1934-8681.
13. Anderson, J., Gee, R., Germane, G., Henry, K. et al., "Analysis of a Real-World High-Speed Rollover Crash from a Video Record and Physical Evidence," SAE Technical Paper 2008-01-1486, 2008, doi:10.4271/2008-01-1486.
14. Asay, A. and Woolley, R., "Rollover Testing on an Actual Highway," *SAE Int. J. Passeng. Cars – Mech. Syst.* 2, no. 1 (2009): 1237-1267, doi:10.4271/2009-01-1544.
15. Stevens, D., Arndt, S., Wayne, L., Arndt, M. et al., "Rollover Crash Test Results: Steer-Induced Rollovers," SAE Technical Paper 2011-01-1114, 2011, doi:10.4271/2011-01-1114.
16. Asay, A., Carter, J., Funk, J., and Stephens, G., "Rollover Testing of a Sport Utility Vehicle (SUV) with an Inertial Measurement Unit (IMU)," SAE Technical Paper 2015 01 1475, 2015, doi:10.4271/2015-01-1475.

CHAPTER 3

17. Larson, R., Smith, J., Werner, S., and Fowler, G., "Vehicle Rollover Testing, Methodologies in Recreating Rollover Collisions," SAE Technical Paper 2000-01-1641, 2000, doi:10.4271/2000-01-1641.
18. Jones, I.S. and Wilson, L.A., "Techniques for the Reconstruction of Rollover Accidents Involving Sport Utility Vehicles, Light Trucks and Minivans," SAE Technical Paper 2000-01-0851, 2000, doi:10.4271/2000-01-0851.
19. Altman, S., Santistevan, D., Hitchings, C., Wallingford, J.G. et al., "Comparison of Rollover Characteristics for Passenger Cars, Light Duty Trucks and Sport Utility Vehicles," SAE Technical Paper 2002-01-0942, 2002, doi:10.4271/2002-01-0942.
20. Herrera, J.M. and Najera, A., "Model for Analyzing Single Vehicle Rollover Accidents," in *Collision—The International Compendium for Crash Research*, Vol. 1, no. 2, ISBN 1934-8681.
21. Rose, N., Beauchamp, G., and Fenton, S., "Factors Influencing Roof-to-Ground Impact Severity: Video Analysis and Analytical Modeling," SAE Technical Paper 2007-01-0726, 2007, doi:10.4271/2007-01-0726.
22. Funk, J., "An Integrated Model of Rolling and Sliding in Rollover Crashes," SAE Technical Paper 2012-01-0605, 2012, doi:10.4271/2012-01-0605.
23. Bready, J., May, A., and Allsop, D., "Physical Evidence Analysis and Roll Velocity Effects in Rollover Accident Reconstruction," SAE Technical Paper 2001-01-1284, 2001, doi:10.4271/2001-01-1284.

Appendix

FIGURE 3.10 Number of rolls versus roll distance, SI units (naturally occurring rollovers, all vehicle types and surfaces).

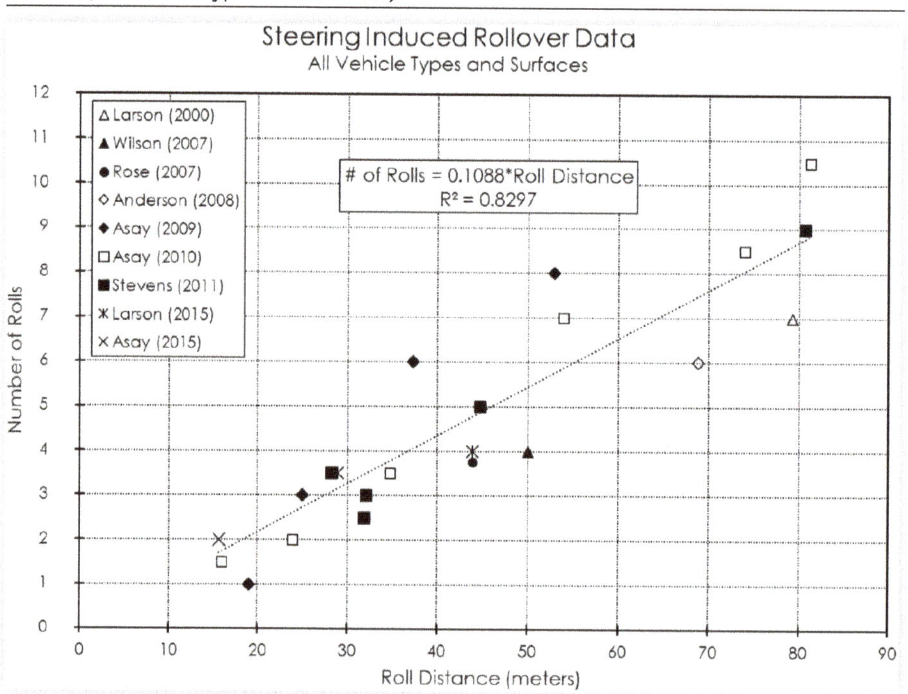

FIGURE 3.11 Number of rolls versus roll distance, SI units (naturally occurring rollovers, SUVs, all surfaces).

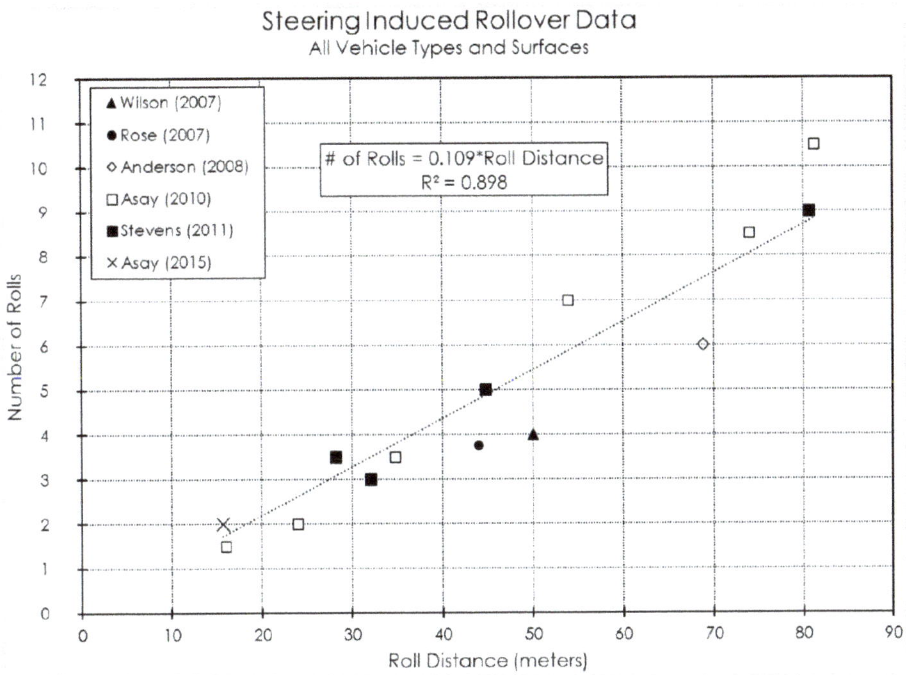

CHAPTER 4

Advanced Analysis Methods: Rollover Phase

Vehicle-to-Ground Impact Model

Impact Model Equations

During a high-speed rollover crash, the vehicle experiences a series of vehicle-to-ground impacts separated by periods of airborne motion. The conditions of these impacts, including the initial velocities; the vehicle's inertial, geometric, and structural properties; and the characteristics of the ground surface, determine the magnitude and direction of the impact forces. These forces, in turn, determine the subsequent motion that the vehicle experiences.

This section develops a planar impact model derived using the principle of impulse and momentum to explore the relationships between the vehicle-to-ground impact conditions and the vehicle's motion [1, 2, 3, 4, 5]. This model is based on the idealized vehicle-to-ground impact shown in Figure 4.1. The vehicle in this figure is depicted in an inverted orientation with the driver's side roof impacting the ground and the front facing into the page. The vehicle has velocity both along and into the ground, and roll velocity that contributes to the speed with which the roof impacts the ground. As a result of this impact, the vehicle is subjected to an impact force that consists of both vertical and ground surface components. The geometry of the impact is defined by the impact radius, which is the distance from the vehicle's center of mass (CoM) to the point at which the impact force is applied, and the impact angle, which is the angle between the ground plane and the impact radius.

Figure 4.2 again depicts the idealized vehicle-to-ground impact of Figure 4.1. However, in this second figure, the descriptive labels of the first have been replaced with

FIGURE 4.1 Idealized vehicle-to-ground impact.

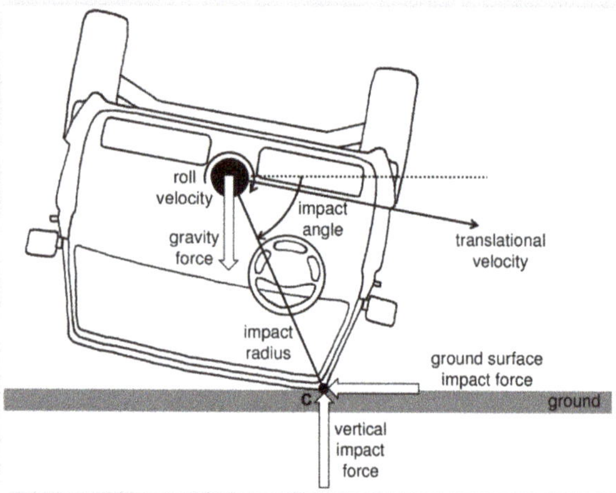

FIGURE 4.2 Impact model symbols.

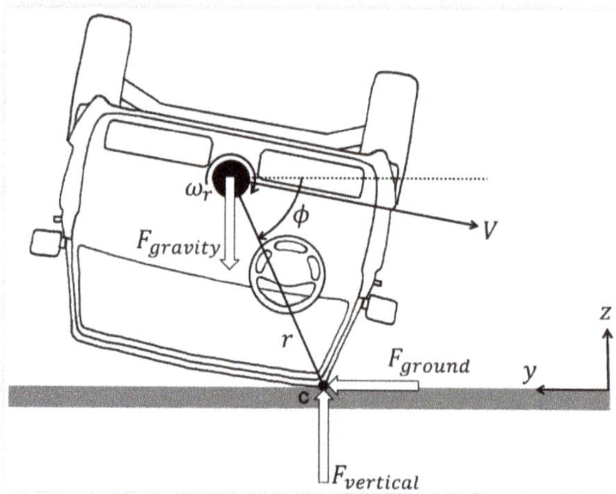

the symbols that the impact model equations utilize. The impact angle and impact radius are designated with the symbols ϕ and r. The velocity vector is designated with the letter v and the vehicle's roll velocity is designated ω_r. During the depicted impact, the vehicle is subjected to both upward and ground surface impact force components, $F_{vertical}$ and F_{ground}, and the gravity force, which is the vehicle's weight, $F_{gravity}$. In general, F_{ground} can act in either the positive (left) or negative (right) direction. On the other hand, $F_{vertical}$ always acts in the positive-z direction and the gravity force always acts in the negative-z direction.

The following assumptions are used in developing equations for this model:

1. The impact is assumed to occur in a single plane (parallel to the surface of the page). Velocity changes along the vehicle's longitudinal axis (in and out of the page) are neglected, as are changes in the pitch and yaw angular velocities.
2. The impact model equations recognize no change in the position of the vehicle through the impact.
3. The impact force is assumed to be concentrated at a single point.
4. It is assumed that no moment arises at the impact point [6].
5. The ground surface impact force is assumed to be a retarding force (frictional) that depends on relative velocity at the contact point for its development. Thus, there can be no velocity reversal (restitution) that occurs in the contact region along the ground plane. Restitution vertical to the ground plane is incorporated into the model.

The principle of impulse and momentum yields the following equations:

$$m\cdot\left(v_{zf}-v_{zi}\right)=P_z-P_g \tag{4.1}$$

$$m\cdot\left(v_{yf}-v_{yi}\right)=P_y \tag{4.2}$$

$$mk_r^2\cdot\left(\omega_{r,f}-\omega_{r,i}\right)=P_y\cdot r\cdot s\phi-P_z\cdot r\cdot c\phi \tag{4.3}$$

In Equations (4.1) through (4.3)

m is the vehicle mass

k_r is the vehicle's radius of gyration for the roll axis (divide the roll moment of inertia by the mass and take the square root)

P_z and P_y are the normal (vertical) and tangential (ground plane) impulse components associated with the vertical and ground plane impact forces

P_g is the gravity impulse

Translational velocity components are denoted with the letter v, final and initial velocities are denoted with the subscripts f and i, and the letters s and c designate the sine and cosine. The gravity impulse can be rewritten with the following equation [7]:

$$P_g = mg \cdot \Delta t \tag{4.4}$$

In Equation (4.4)

g is the gravitational constant

Δt is the impact duration

The following constraint equations govern the impact energy loss along the normal and tangential directions:

$$e = -\frac{v_{zf} - r \cdot c\phi \cdot \omega_{rf}}{v_{zi} - r \cdot c\phi \cdot \omega_{ri}} \tag{4.5}$$

$$\mu = \frac{P_y}{P_z} \tag{4.6}$$

In Equation (4.5), e is the coefficient of restitution for the impact, which is defined as the negative ratio of the post-impact to pre-impact vertical velocities at the point of collision force transfer (Point C). In Equation (4.6), μ is the impulse ratio, which establishes the magnitude of the tangential impulse relative to the magnitude of the normal impulse. In many cases, the impulse ratio can be thought of as a coulomb friction value, though its application is not limited to this interpretation [8, 9, 10, 11, 12]. In addition to the effects of friction between the ground and the vehicle body, the ground plane impulse may also include the effects of forces generated by snagging between the vehicle and the ground or furrowing of the vehicle into the ground. The "available friction" can be set at a value that reflects such snagging or furrowing when it occurs.

Substituting Equations (4.1), (4.2), and (4.4) into Equation (4.6) yields the following equation:

$$v_{yf} = v_{yi} + \mu \cdot \left[\left(v_{zf} - v_{zi}\right) + g\Delta t\right] \tag{4.7}$$

Substituting Equation (4.6) into Equation (4.3) yields the following equation:

$$mk_r^2 \cdot \left(\omega_{r,f} - \omega_{r,i}\right) = P_z \cdot r \cdot \left(\mu \cdot s\phi - c\phi\right) \tag{4.8}$$

Substituting Equations (4.1) and (4.4) into Equation (4.8) yields the following equation:

$$\omega_{rf} = \omega_{ri} + \left[\left(v_{zf} - v_{zi}\right) + g\Delta t\right] \cdot \frac{r \cdot \left(\mu \cdot s\phi - c\phi\right)}{k_r^2} \tag{4.9}$$

Now, rearranging Equation (4.5) to solve for the final roll velocity yields the following equation:

$$\omega_{rf} = \frac{1}{r \cdot c\phi} \cdot v_{zf} + \frac{e}{r \cdot c\phi} \cdot \left(v_{zi} - r \cdot c\phi \cdot \omega_{ri}\right) \tag{4.10}$$

Next, equating Equations (4.9) and (4.10) through the final roll velocity yields the following equation:

$$v_{zf} = v_{zi} - (1+e)v_{zc,i}\left\{\frac{k_r^2}{k_r^2 - r^2 c\phi(\mu \cdot s\phi - c\phi)}\right\} + g\Delta t_i\left\{\frac{r^2 c\phi(\mu \cdot s\phi - c\phi)}{k_r^2 - r^2 c\phi(\mu \cdot s\phi - c\phi)}\right\} \tag{4.11}$$

In this equation, $v_{zc,i}$ is the vertical velocity of the vehicle at Point C immediately preceding the ground contact. This velocity is obtained with the following equation, which is related to the vehicle's CoM vertical velocity (v_{zi}), its roll velocity (ω_{ri}), and the impact angle and radius:

$$v_{zc,i} = v_{zi} - r \cdot c\phi \cdot \omega_{ri} \tag{4.12}$$

So, Equations (4.7), (4.9), and (4.11) yield the final velocities or the CoM velocity changes experienced by the vehicle during a vehicle-to-ground impact given the initial velocities and the impact geometry. When using these equations, the analyst would solve Equation (4.11) first, then the result would feed into Equations (4.7) and (4.9).

Impulse Ratio (μ)

Within these equations, the sign of the impulse ratio governs the direction of the ground plane collision force and also affects the magnitude of the vertical and roll velocity changes. A positive impulse ratio produces a ground plane force that acts in the positive direction, and a negative impulse ratio results in a ground plane force that acts in the negative direction. The direction of the ground plane impact force, in turn, determines whether the vehicle experiences a positive or negative ground plane velocity change and whether the ground surface impact force tends to increase or decrease the roll velocity. For the passenger side leading roll depicted in the images of this section, a positive F_{ground} acts to accelerate the roll and a negative F_{ground} acts to decelerate the roll. This trend is reversed for a driver side leading roll.

Physically, the ground plane impulse acts in a direction opposing the velocity of the vehicle in its region of contact with the ground (Figure 4.3) [13]. Thus, the impulse ratio can be further specified with the following equation:

$$\mu = -\mu_0 \cdot sgn(v_{yc,i}) \tag{4.13}$$

FIGURE 4.3 Direction of ground plane impulse.

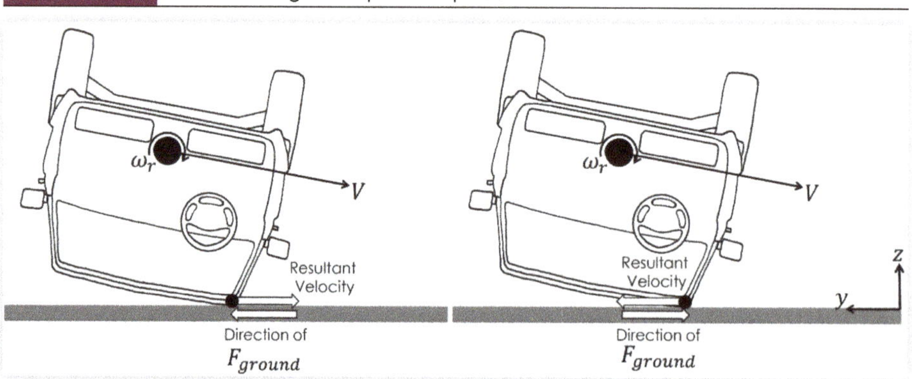

In this equation,

μ_0 is the nominal value of the impulse ratio that defines the magnitude of the ground plane impulse relative to the vertical impulse

$v_{yc,i}$ is the ground plane velocity of the vehicle at Point C

This velocity is given by the following equation:

$$v_{yc,i} = v_{yi} + r \cdot s\phi \cdot \omega_{ri} \tag{4.14}$$

Thus, for a particular vehicle-to-ground impact, the relative magnitude of the vehicle's CoM ground plane velocity (v_{yi}) and the ground plane component of its rotational perimeter velocity ($r \cdot s\phi \cdot \omega_{ri}$) determines the sign of $v_{yc,i}$ and, thus, the sign of μ.

For any particular vehicle-to-ground impact, there is a critical value of the impulse ratio, μ_c, which causes relative motion between the vehicle and the ground to cease in the contact region. The impulse ratio will not take on a value that exceeds this critical impulse ratio. Therefore, the magnitude of the available friction coefficient relative to the magnitude of the critical impulse ratio for a particular impact has physical significance.

The *available* friction coefficient represents the magnitude of friction force that *can be* recruited during the impact. The *critical* impulse ratio represents the magnitude of friction force that must be recruited in order for relative motion to cease along the ground surface in the contact region between the vehicle and the ground. When the critical impulse ratio is greater than the available friction coefficient, then all of the available friction is recruited during the impact, but that friction is insufficient to cause sliding to cease in the contact region. When the available friction coefficient exceeds the critical impulse ratio, only a portion of the available friction is recruited and sliding ceases in the contact region. In such cases, the value of the impulse ratio for the impact model should be set at the value of the critical impulse ratio, not the available friction coefficient. This is because recruitment of the available friction depends on relative velocity being present between the ground and the vehicle body. Once this relative motion ceases, no additional friction can be recruited. These concepts can be summarized as follows:

$$\mu_c > \mu_{available} \rightarrow \mu = \mu_{available}$$

$$\mu_c < \mu_{available} \rightarrow \mu = \mu_c$$

The critical impulse ratio for any particular vehicle-to-ground impact is the value of the impulse ratio that results in the final velocity at the Point C being zero, as follows:

$$v_{yc,f} = v_{y,f} + r \cdot s\phi \cdot \omega_{rf} = 0 \tag{4.15}$$

Equation (4.16) is an analytical expression that yields the impulse ratio that results in the condition of Equation (4.15) being satisfied. Rose [14] details the derivation of Equation (4.16) and it is not covered here.

$$\mu_c = \frac{\left[(1+e)\cdot v_{zc,i} - g\Delta t\right]\cdot\left[r^2 \cdot s\phi \cdot c\phi\right] + v_{yc,i}\cdot\left[k_r^2 + r^2 \cdot c^2\phi\right]}{\left[(1+e)\cdot v_{zc,i} - g\Delta t\right]\cdot\left[k_r^2 \cdot r^2 \cdot s^2\phi\right] + v_{yc,i}\cdot\left[r^2 + s\phi^2 \cdot c\phi\right]} \tag{4.16}$$

As this equation shows, the critical impulse ratio depends on the following factors: (1) the vehicle's initial velocity conditions for the impact, (2) the coefficient of restitution, (3) the impact duration, and (4) the impact configuration and geometry (which can include the deformed shape of the vehicle).

Figure 4.4 is a graph showing the value of the critical impulse ratio, calculated with Equation (4.16), for a number of impact scenarios. In this graph, the vehicle's over-the-ground (OTG) speed is plotted on the horizontal axis and the critical impulse ratio (or available coefficient of friction) on the vertical axis. The impact scenarios depicted in this graph are for a vehicle's CoM downward speed of 2 mph (3.2 kph), a roll velocity of

CHAPTER 4

FIGURE 4.4 Critical impulse ratio for a number of impact scenarios.

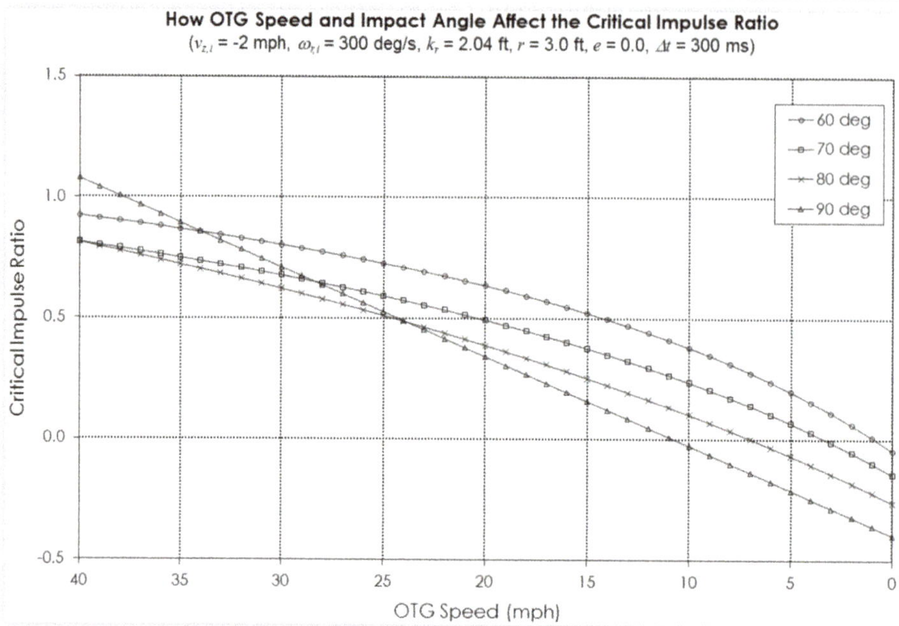

300°/s, a radius of gyration for the roll axis of 2.04 ft (0.62 m), an impact radius of 3.0 ft (0.91 m), a coefficient of restitution of 0.0, and an impact duration of 300 ms. The graph contains curves for impact angles of 60, 70, 80 and 90°. A version of this graph containing SI units is included in the appendix to this chapter.

This graph demonstrates that for a fixed roll velocity and impact configuration, the critical impulse ratio decreases as the vehicle's OTG speed decreases. For the sake of illustration, consider a situation in which the available friction coefficient can be assumed to be 0.5. If this is the case, then Figure 4.5 shows the value which the impulse ratio

FIGURE 4.5 Impulse ratio for a number of impact scenarios.

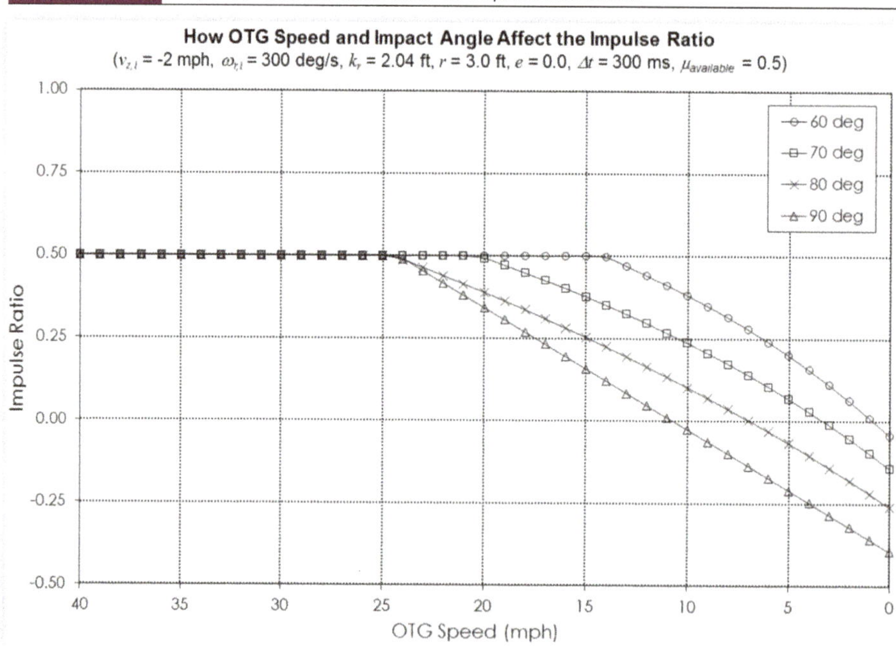

FIGURE 4.6 Dependence of impact angle on deformation.

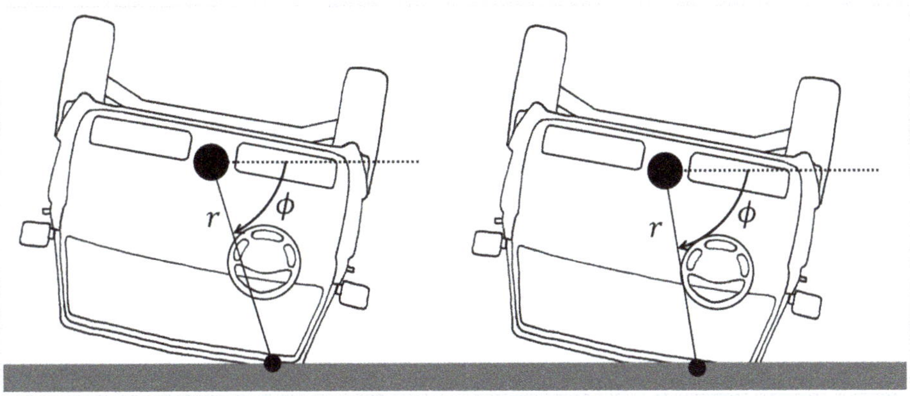

would take on for each of the impact scenarios of Figure 4.4. For all of the impact scenarios occurring at OTG speeds above 25 mph (40.3 kph), the critical impulse ratio exceeds the available friction coefficient. Thus, for these scenarios, the impulse ratio takes on a value of 0.5. As the OTG speed drops below 25 mph (40.3 kph), impact scenarios begin to arise in which the critical impulse ratios are less than the available friction. In these cases, the impulse ratio takes on the value of the critical impulse ratio.

At this point, it can be observed that even if the impulse ratio is limited in accordance with Equation (4.16), physically unrealistic solutions that produce an increase in the total energy of the vehicle are still mathematically possible [11]. Thus, the analyst can also impose the constraint that the vehicle's total energy cannot increase during an impact. If a solution is obtained that generates an increase in the vehicle's energy, then the analyst can further reduce the impulse ratio until this condition ceases to exist. The next equation below yields the energy loss experienced by the vehicle during a vehicle-to-ground impact. In this equation, the symbol T_L has been used to designate the energy loss.

$$T_L = \frac{1}{2} m \cdot \left(v_{yi}^2 - v_{yf}^2 + v_{zi}^2 - v_{zf}^2 + k_r^2 \omega_{ri}^2 - k_r^2 \omega_{rf}^2 \right) \tag{4.17}$$

Impact Angle (ϕ)

Now consider the impact angle (ϕ). The value of this parameter for any particular impact depends on a number of factors, including the roll orientation of the vehicle, the shape of the vehicle, and the amount of deformation that occurs during the impact. To see how this parameter depends on the amount of deformation, consider Figure 4.6. This figure depicts two roof-to-ground impacts that occur with identical roll angles, but with roof deformation magnitudes that vary (represented by how much of the roof is pressed against the ground). The roof deformation for the impact on the right is greater than for the impact on the left. For each of these impacts, the Point C has been positioned approximately at the center of the deformation region. In this case, increasing deformation has resulted in an increasing impact angle and a decreasing impact radius.

Nonconstant Deceleration Models

The rate at which a rolling vehicle decelerates is dictated by the accumulation of the ground plane velocity changes that occur during the rollover. Thus, any factor that influences the ground surface velocity changes could influence the deceleration rate that the vehicle experiences over the roll distance. These factors include the following: (1) the available surface friction; (2) OTG speeds, vertical velocities, and roll velocities experienced during the roll; (3) the orientations of the specific vehicle-to-ground impacts; (4) the vehicle geometry; and (5) the stiffness of the vehicle structures engaged during the roll.

Rather than attempting to account for all of these factors on an impact-by-impact basis, reconstructionists have often assumed that a rolling vehicle decelerates at a constant rate. It is obvious that this is not actually the case. Carter [15] examined the accuracy of using a constant deceleration rate for rollover reconstruction by comparing crash test data to reconstructed results. For the two rollover crash tests that he examined, Carter found that the vehicles' deceleration rates were not constant over the roll distance and that they were higher during the early portions of the rolls than during the latter. Thus, when using a constant deceleration rate approach, the deceleration rates for the early portions of the rolls were underestimated, and those during the later portions were overestimated. For the two cases examined by Carter, the constant deceleration approach overestimated the vehicles' OTG speeds at most discrete points along the roll distances.

Estimates of a vehicle's roll velocity history are obtained by determining roll positions at discrete points along the roll path, and then estimating the time that elapsed while the vehicle traversed the distance between these positions. Since the deceleration rate influences these time estimates, it also influences the roll velocities that are obtained. Thus, using the constant deceleration approach can also introduce errors into estimates of the roll velocity history. Carter found that the accuracy of the constant deceleration approach depended on the number of roll positions that could be established during the roll phase, but in general, the constant deceleration approach tended to overestimate the peak roll velocities. Calculations of a vehicle's instantaneous OTG and roll velocities could be improved by allowing for the vehicle's deceleration rate to vary along the roll distance in a realistic fashion. The accuracy of these instantaneous velocities becomes relevant and significant when one begins to consider a rollover on an event-by-event basis, considering each ground impact and airborne phase as a discrete event, and when one uses mathematical models to analyze occupant dynamics.

There are generally two approaches that have been employed in the accident reconstruction literature for introducing a variable deceleration rate into analysis of the roll phase. The first approach has been termed the "discrete regions approach." In this approach, the analyst would use a constant deceleration rate approach, but vary that deceleration rate in a sequence of regions, as depicted in Figure 4.7. This graph depicts a rollover consisting of three regions, each having a different average deceleration rate.

FIGURE 4.7 Discrete regions approach.

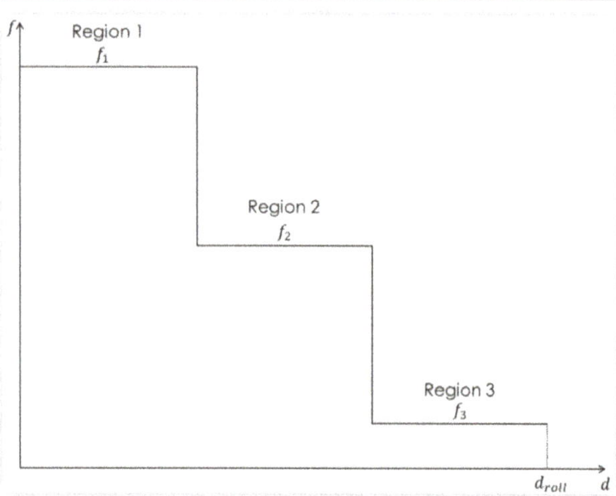

To use this approach, one would need to decide how many regions to use and what criteria to employ in identifying and separating the regions. Part of the discussion in this section focuses on a rationale for identifying such regions. This discussion shows that a vehicle's roll velocity history reveals regions in which the vehicle's deceleration rate would vary.

To develop this first approach, consider Equation (4.18), which yields the change in roll velocity for a vehicle-to-ground impact:

$$\Delta\omega_r = \left(\Delta V_z + g\Delta t\right)\cdot\frac{r\cdot\left(\mu\cdot s\phi - c\phi\right)}{k_r^2} \tag{4.18}$$

In this equation, the sign of the $\mu \cdot s\phi - c\phi$ term varies with the impulse ratio and the impact angle and determines whether the vehicle's roll velocity increases or decreases. The other quantities in this equation are always positive. Considering the three-region shape of the typical roll velocity history discussed in Chapter 3, this $\mu \cdot s\phi - c\phi$ term tends to be positive early in the roll phase at higher OTG speeds, near zero in the area of the roll velocity plateau, and then negative later in the roll phase as the OTG speed continues to decrease and the roll velocity begins decreasing.

Similarly, the following equation yields a vehicle's average OTG deceleration rate for a particular vehicle-to-ground impact:

$$f_{impact} = \left[\frac{k_r^2}{r}\right] \cdot \left[\frac{\mu}{\mu \cdot s\phi - c\phi}\right] \cdot \left[\frac{\Delta\omega_r}{g\Delta t}\right] \tag{4.19}$$

In this equation, f_{impact} is the average ground plane deceleration rate for the vehicle during the vehicle-to-ground impact. To obtain Equation (4.19), Equation (4.9) was substituted into Equation (4.7) and the result was divided through by $g\Delta t$. This equation also contains the $\mu \cdot s\phi - c\phi$ term. This equation reveals the dependence of f_{impact} on the impulse ratio, the impact configuration, the roll inertia of the vehicle, the change in roll velocity, and the impact duration. This equation is significant for understanding how a vehicle's deceleration rate would vary over the course of a rollover because it reveals a relationship between the change in roll velocity experienced by the vehicle during a ground impact and the deceleration rate for that impact. This indicates a relationship between the vehicle's roll velocity history and its OTG deceleration rate history. Given this relationship, the discrete regions approach could use a different constant deceleration rate for each region of the roll velocity curve.

Figure 4.8 is a graph showing a number of vehicle-to-ground impact scenarios with the change in roll velocity plotted on the horizontal axis, and the vehicle's average deceleration rate for the impact plotted on the vertical axis. The horizontal axis progresses from large increases in roll velocity to large decreases in roll velocity. Interpreted in the context of a three-region roll velocity history, we can observe that Region 1 would encompass the left portion of the graph, Region 2 the middle portion, and Region 3 the right portion. The graph contains curves for four different impact angles with impact scenarios having the following conditions: (1) downward CoM velocity, 2 mph (3.2 kph); (2) initial

FIGURE 4.8 Relationship between roll velocity changes and the deceleration rate.

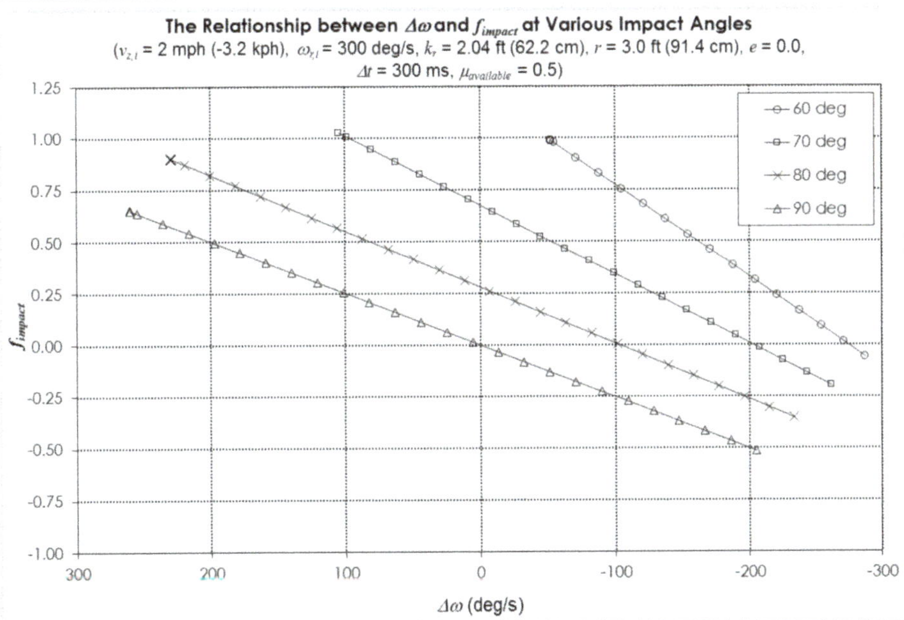

FIGURE 4.9 Continuously variable approach.

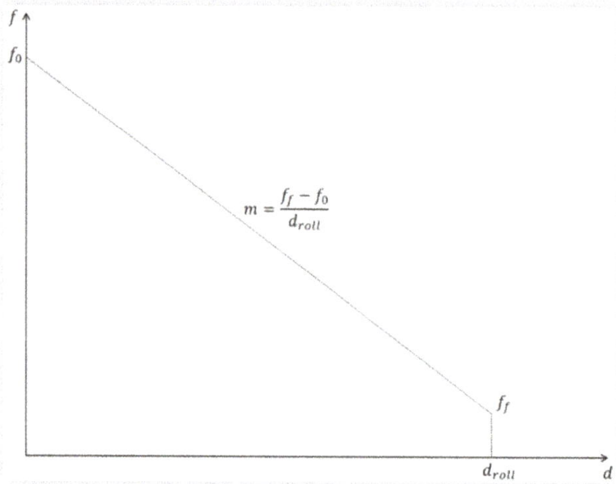

roll velocity, 300°/s; (3) radius of gyration for the roll axis, 2.04 ft (0.62 m); (4) impact radius, 3.0 ft (0.91 m); (5) coefficient of restitution, 0.00; (6) impact duration, 300 ms; and (7) available friction coefficient, 0.5. For all of the impact angles depicted, the average deceleration rates associated with the impact scenarios diminishes as one moves from left to right on the graph. Again interpreted in the context of the typical three-region roll velocity history, this would imply that the deceleration rates associated with impacts would decrease as the vehicle progresses through a three-region roll velocity history. Thus, the deceleration rate would start at its highest level and steadily diminish through the roll phase.

Given the characteristics of the curves in Figure 4.8, it might also be reasonable to vary the deceleration rate during the roll phase in a continuous manner. Thus, the second non-constant deceleration rate approach that has been employed in the accident reconstruction literature has been termed the "continuously variable approach." This approach involves specifying the variability of the deceleration rate over the course of a rollover with a continuous function. The simplest assumption for this approach would be to assume the deceleration rate decreases linearly with the roll distance as shown in Figure 4.9. With this assumption, the analyst would either specify the initial and final deceleration rates or the initial deceleration rate and the rate at which it decreases.

Validation of a Nonconstant Deceleration Model

This section explores the degree to which using a variable deceleration rate actually improves the accuracy of a reconstruction. Rose and Beauchamp [14], Rose and Beauchamp [16] explored this question by using various variable deceleration rate profiles to analyze the two dolly rollover crash tests reported by Luepke in 2007 [17] and then further analyzed by both Luepke and Carter in 2008 [18]. Both tests were dolly rollover tests where the vehicles were launched onto a 6″ (15 cm) layer of lightly compacted desert soil. In the first test, run with a 1998 Ford Expedition, the test vehicle was launched at a speed of 43.2 mph (69.5 kph). From the time it exited the dolly until the time it came to rest, the vehicle traveled approximately 120 ft (36.6 m) and rolled 4 times. The average deceleration rate over this distance (trip and roll phases combined) was 0.52g. In the second test, run with a 2004 Volvo XC90, the test vehicle was launched at a speed of 42.9 mph (69.0 kph). From the time it exited the dolly until the time it came to rest, the vehicle traveled approximately 115 ft (35.1 m) and rolled 4¼ times. The average deceleration rate over this distance (trip and roll phases combined) was 0.53g.

Carter conducted a frame-by-frame analysis of the video footage of these tests and correlated the vehicle positions and orientations to the physical evidence deposited on the test surface. Based on the positions of the vehicles in these tests, Carter calculated OTG speed versus time curves for each of the tests. He obtained the roll velocity curves for these tests from sensor data. The analysis by Rose and Beauchamp relied on this analysis previously conducted by Carter.

In their 2009 study, Rose and Beauchamp considered four deceleration rate models in their analysis - a constant deceleration rate; a step function deceleration rate profile in which the deceleration rate leading up to peak roll velocity was set at a constant value higher than the average, and the deceleration rate following the peak roll velocity was set at a value lower than the average; a linearly decreasing deceleration rate profile in which the deceleration rate began high and steadily progressed to a low value; and finally, a function that began with a high and constant deceleration rate, and then at the time of the peak roll velocity, began to decrease linearly. Rose and Beauchamp concluded that the linearly decreasing deceleration rate profile was the easiest to implement and that it yielded significant improvements in the calculated translational and angular velocity time histories when compared to the constant deceleration rate model.

Figure 4.10 shows the actual speed curve for the Expedition test, along with speed curves obtained with the constant deceleration rate model and the linearly decreasing deceleration rate model (a version of this graph containing SI units is included in the Appendix to this chapter). Figure 4.11 shows the actual roll velocity curve for the Expedition test, along with the roll velocity curves obtained with the constant deceleration rate model and the linearly decreasing deceleration rate model. Figures 4.12 and 4.13 show similar results for the XC90 test (a version of Figure 4.12 containing SI units is included in the Appendix). The data presented in these graphs demonstrates that, for both of these tests, the linearly decreasing deceleration rate better predicted the actual roll phase duration, the actual decrease in speed over the course of the roll, and the actual peak roll velocity.

In their 2010 study, Rose and Beauchamp further developed the linearly decreasing deceleration rate approach, presenting equations for the model, presenting a step-by-step procedure, and again demonstrating that the technique led to a significantly improved match between a reconstruction and the known OTG speed, roll velocity, and roll duration. They considered linearly decreasing deceleration rate profiles that yielded average deceleration rates in the range of 0.4 to 0.6g. The results using this range tended to bracket the actual duration of the roll phase, the actual initial speed, and the actual peak roll velocity. Both ends of the range significantly improved the results over the constant deceleration rate approach.

Arndt [19] applied the linearly decreasing deceleration rate approach to the steering-induced rollovers reported by Stevens [20]. Arndt compared the known OTG speeds

FIGURE 4.10 Comparison of OTG speed obtained with constant deceleration and linearly decreasing deceleration models - Expedition.

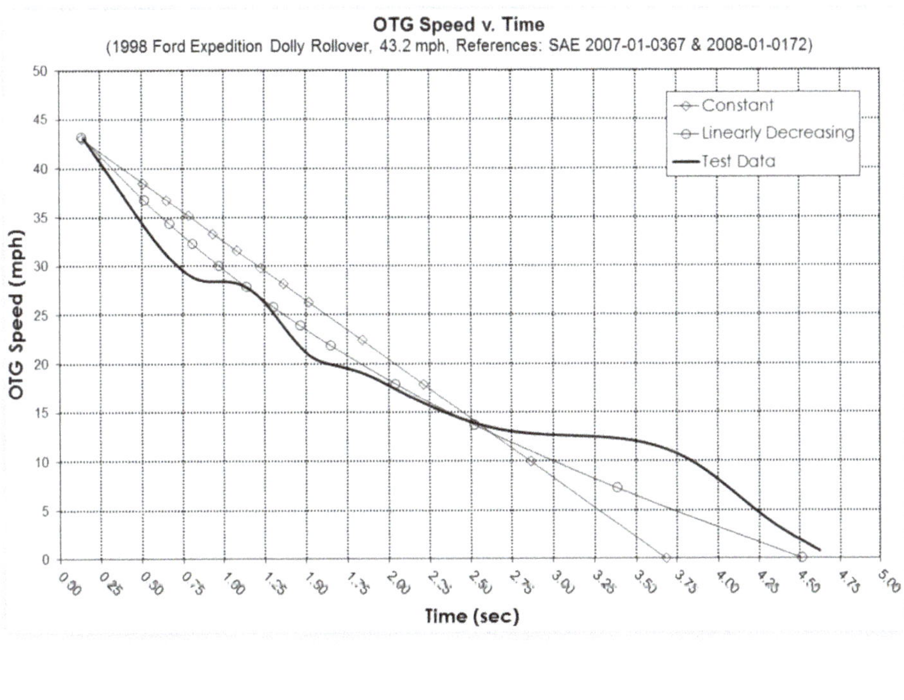

FIGURE 4.11 Comparison of roll velocity obtained with constant deceleration model and linearly decreasing deceleration model - Expedition.

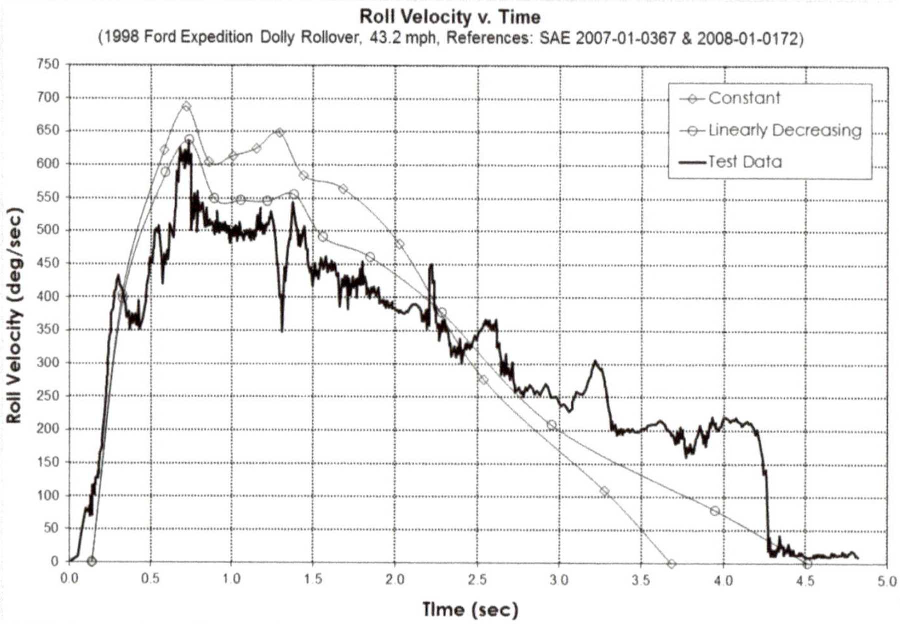

FIGURE 4.12 Comparison of OTG speed obtained with constant deceleration model and linearly decreasing deceleration model - XC90.

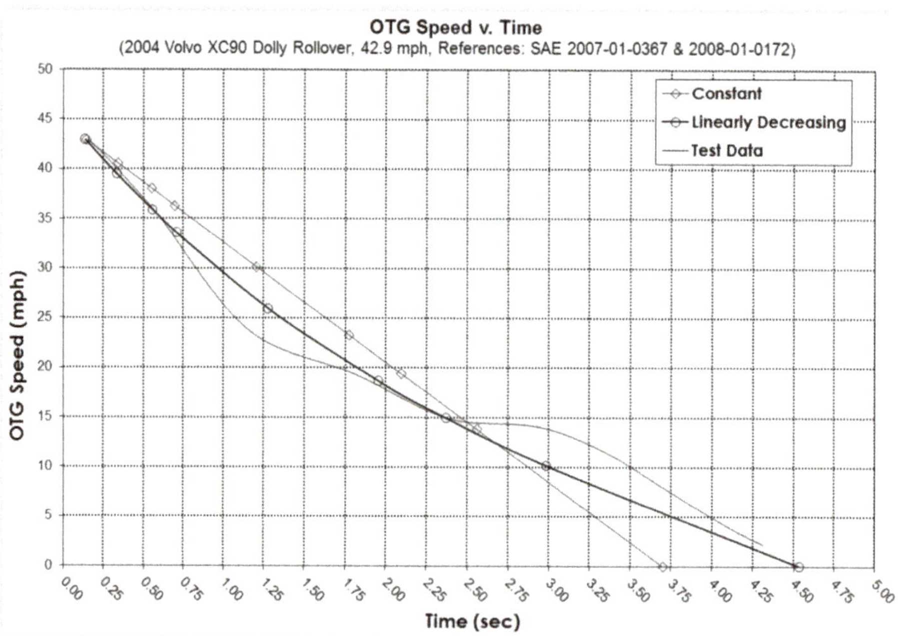

FIGURE 4.13 Comparison of roll velocity obtained with constant deceleration model and linearly decreasing deceleration model - XC90.

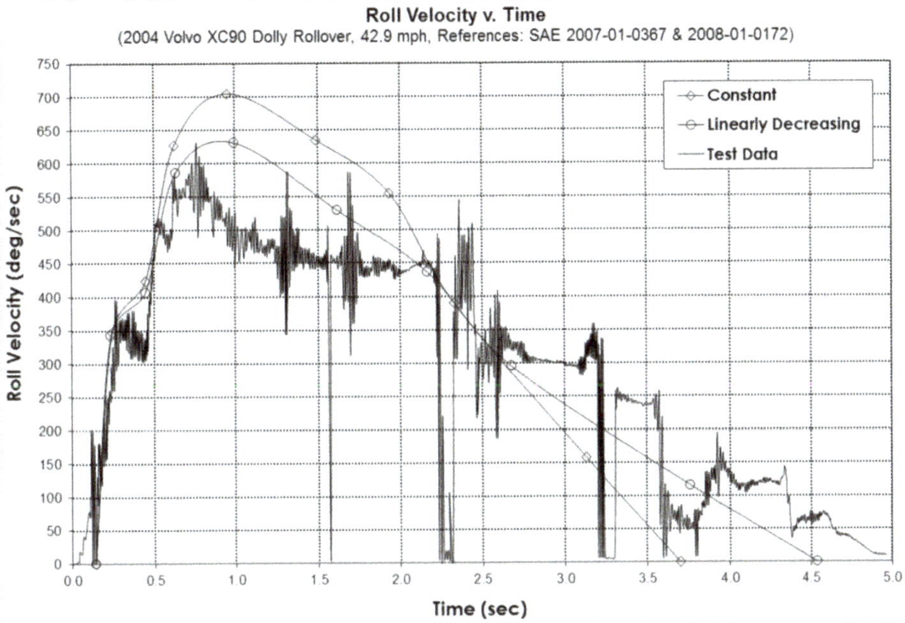

and roll rates from these five tests to the speeds and roll rates obtained from a reconstruction utilizing the linearly decreasing deceleration rate approach. Arndt found that linearly decreasing deceleration rate model yielded reasonable agreement with the measured speeds, roll velocities, and rollover duration from these naturally occurring rollovers, and that the method significantly improved the accuracy of the speed and roll rate time histories when compared to a method using a constant deceleration rate.

One interesting aspect of Arndt's methodology was that he began applying the linearly decreasing deceleration rate "from first post-trip ground contact" rather than from the beginning of the roll phase. For each test, Arndt first constructed a deceleration rate profile where the initial deceleration rate was 0.2g higher than the average deceleration rate for the test, and the ending deceleration rate was 0.2g lower than the average. For each test, he then also considered a low-end deceleration rate where the average deceleration rate was 0.38g and a high-end deceleration rate where the average deceleration rate was 0.5g. For each of these profiles, he again set the initial deceleration rate to be 0.2g higher than the average and the ending deceleration rate was 0.2g lower than the average.

In describing his results, Arndt made two significant observations. First, "a key factor in predicting the roll rate time history is choosing a deceleration model that properly predicts the total rollover duration. A model that tends to underestimate the rollover duration...will overestimate the roll rate. Conversely, a model that overestimates the rollover duration will underestimate roll rates." Second, "One of the shortcomings of the [linearly decreasing deceleration rate] approach in modeling the Stevens roll tests was that it assumed a constant rate of decreasing deceleration with the average deceleration at the midpoint of the rollover distance. The Stevens rollover testing consistently showed that the maximum roll rate and an associated drop in deceleration rate occur in the first half of the rollover distance. The effect of using the [linearly decreasing deceleration rate approach] in modeling the Stevens test was to delay the predicted time of the maximum roll rate. A bi-constant deceleration approach...with a change in deceleration rate at the point in the rollover when the [over the ground speed] versus time curve changes slope, may more closely predict the time of the maximum roll rate." A bi-constant deceleration rate approach is equivalent to the discrete regions approach described earlier.

Funk [21] presented a bi-constant deceleration approach similar to what Arndt proposed. Funk referred to his model as pseudo-cylindrical since it neglected impacts (like a rolling cylinder) but still allowed for the vertical force at the contact to occur ahead of the CoM (thus, roll deceleration could be modeled). In describing his model, Funk stated it was based on the "hypothesis that the tumbling phase of a rollover consists of three distinct phases: an initial brief airborne phase between rollover initiation and the first ground contact, an early phase in which the vehicle rolls while also sliding along the ground, and a later phase in which the vehicle rolls without sliding." Based on his model, Funk noted that "the predicted translational deceleration is higher in the first phase of the rollover than in the second phase. The predicted roll velocity is a triangular-shaped function in which the peak roll rate occurs at the transition between the first and second phases of the rollover." This model fits well with a number of rollover crash tests that Funk examined and his article reported model parameters for 12 tests.

Reconstruction Methodology: Roll Phase (Linearly Decreasing Deceleration)

Equation (4.20) yields the average deceleration (f_{avg}) given the initial (f_0) and final (f_f) deceleration rates. Given the results of the Rose and Beauchamp and the Arndt [19, 22] studies, one could perhaps set the average deceleration rate at 0.44g, the initial deceleration rate at 0.64g, and the final deceleration rate at 0.24g.

$$f_{avg} = \frac{f_0 + f_f}{2} \tag{4.20}$$

The slope (m) of the linearly decreasing deceleration rate profile is then given by the following equation:

$$m = \frac{f_f - f_0}{d_{roll}} \tag{4.21}$$

And thus, the equation for the linearly decreasing deceleration rate is as follows:

$$f_d = f_0 + \frac{f_f - f_0}{d_{roll}} \cdot d \tag{4.22}$$

In Equation (4.22), d is an arbitrary distance along the roll path, with zero distance starting at the beginning of the roll (or at the first ground impact). The variable f_d is the instantaneous drag factor at this arbitrary distance. Now, the deceleration rate at the distance d can be obtained as a function of the vehicle's velocity with Equation (4.23).

$$a_d = \frac{dv_d}{dt} = \frac{dv_d}{dd} \cdot \frac{dd}{dt} = v_d \cdot \frac{dv_d}{dd} = -f_d g \tag{4.23}$$

Equation (4.23) can be integrated as shown in Equation (4.24). The left side of this equation is integrated from the distance d to the end of the roll phase. The right side is integrated from the vehicle's OTG speed at the distance d to the speed at rest, which is zero.

$$-\int_{d}^{d_{roll}}\left[f_0+\frac{f_f-f_0}{d_{roll}}\cdot d\right]\cdot g\cdot dd=\int_{v_d}^{0}v_d\cdot dv_d \quad (4.24)$$

Integrating Equation (4.24) yields the following equation:

$$-g\left[f_0\cdot d+\frac{f_f-f_0}{2\cdot d_{roll}}\cdot d^2\right]_{d}^{d_{roll}}=\left[\frac{v_d^2}{2}\right]_{v_d}^{0} \quad (4.25)$$

Applying the extents of the integration and rearranging algebraically yields the following:

$$g\cdot\left[f_0\left(d_{roll}-d\right)-\frac{f_0-f_f}{2\cdot d_{roll}}\cdot\left(d_{roll}^2-d^2\right)\right]=\frac{v_d^2}{2} \quad (4.26)$$

Solving for the velocity of the vehicle at the distance d yields Equation (4.27):

$$v_d=\sqrt{2\cdot g\cdot f_0\cdot\left(d_{roll}-d\right)-g\cdot\frac{f_0-f_f}{d_{roll}}\cdot\left(d_{roll}^2-d^2\right)} \quad (4.27)$$

To quantify the time that elapses between any two positions along the roll path, d_i and d_{i+1}, the vehicles average velocity ($v_{avg,\, i\rightarrow i+1}$) between these positions needs to be calculated. This average velocity can be calculated by subdividing the distance between d_i and d_{i+1} into small intervals, using a spreadsheet to calculate the velocity at each of these subdivisions, then summing these and dividing by the number of values to obtain the average. Once this average velocity is known, then the time that elapses between the two positions can by calculated with the following equation:

$$\Delta t_{i\rightarrow i+1}=\frac{\Delta d_{i\rightarrow i+1}}{v_{avg,i\rightarrow i+1}} \quad (4.28)$$

Once this time is calculated, the vehicle's average roll velocity for the segment can be calculated with the following expression:

$$\omega_{avg,i\rightarrow i+1}=\frac{\theta_{i+1}-\theta_i}{\Delta t_{i\rightarrow i+1}} \quad (4.29)$$

In this equation, θ_i and θ_{i+1} are the roll angles at positions i and $i+1$. These equations in place, the following step-by-step methodology can be used to implement a linearly decreasing deceleration profile within the reconstruction of the roll phase:

1. **Map the scene and vehicle evidence.** Produce a scaled scene diagram containing the physical evidence from the rollover and a computer model of the vehicle.
2. **Measure the roll distance.** This is the distance from the CoM at the neutral stability position (when the vehicle goes airborne) to the CoM of the vehicle at the rest position.

3. **Determine positions and orientations for the vehicle along the roll path.** Measure the distances between these roll positions.
4. **Assign values for the initial and final deceleration rates for the roll phase.** A reasonable midrange assumption would be an initial value of 0.64g and a final value of 0.24g. This would yield an average value of 0.44g.
5. **Calculate the velocity-distance history for the vehicle** from the beginning of the roll distance to rest using Equation (4.27).
6. **Calculate the elapsed time between each set of reconstructed vehicle positions** with Equation (4.28).
7. **Calculate the segment roll velocities** with Equation (4.29). This requires that the roll angle be measured at each reconstructed vehicle position along the roll path. The roll angle should be measured cumulatively, meaning that at the completion of one roll the roll angle would be 360°, at the completion of two rolls the roll angle would be 720°, and so on. Also, calculate the number of rolls accumulated at each position, which would be the roll angle divided by 360.
8. Make adjustments to the reconstructed roll positions, ensuring that they remain consistent with the physical evidence, but also produce a roll velocity curve that is consistent with the trends discussed in Chapter 3.

Case Study: Asay's Isuzu Rodeo Rollover Test

Consider again the Isuzu Rodeo rollover test reported by Asay [23]. This vehicle rolled 7 times in approximately 181 ft (55.3 m). The physical evidence was diagrammed and the roll positions depicted in Figure 4.14 were determined from this evidence. Figure 4.14 is identical to Figure 3.7. The segment distances and roll angles associated with these roll positions are listed in Table 4.1.

Analysis is carried out here using both an average deceleration rate of 0.363g (actual) and an average deceleration rate of 0.44g (a midrange value likely to be employed by a reconstructionist). For an average deceleration rate of 0.363g, the initial deceleration rate is set at 0.563g and the ending deceleration rate at 0.163g. For an average deceleration rate of 0.44g, the initial deceleration rate is set at 0.64g and the ending deceleration rate

FIGURE 4.14 Roll positions determined from the scene evidence.

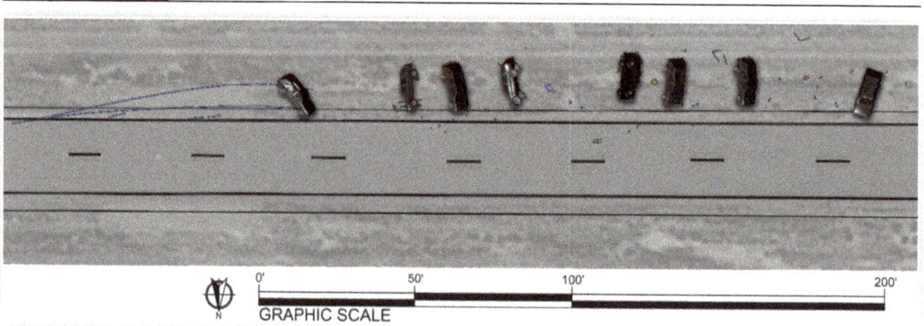

TABLE 4.1 Roll phase calculations for Asay's Isuzu Rodeo test

Segment #	Segment distance		Roll angle (°)	Vehicle speed (mph/kph)		Segment duration (s)	Segment roll velocity (°/s)
End of trip			48	44.4	71.5		
1	35.4 ft	10.8 m				0.59	378
			270	37.7	60.7		
2	14.4 ft	4.4 m				0.27	462
			395	34.9	56.2		
3	18.5 ft	5.6 m				0.38	524
			595	31.2	50.2		
4	36.2 ft	11.0 m				0.90	681
			1205	23.9	38.5		
5	14.9 ft	4.5 m				0.46	582
			1470	20.7	33.3		
6	22.9 ft	7.0 m				0.86	435
			1845	15.5	24.9		
7	39.0 ft	11.9 m				3.41	198
Rest			2520	0			
Totals	181.3 ft	55.3 m				6.88	

is set at 0.24g. Equation (4.27) can be used to calculate the speed of the vehicle at each of the positions depicted in Figure 4.14. As an example, substituting the total roll distance and the deceleration rates for the 0.363g scenario into this equation yields the following:

$$v_d = \sqrt{2 \cdot 32.2 \cdot 0.563 \cdot (181.3 - d) - 32.2 \cdot \frac{0.563 - 0.163}{181.3} \cdot (181.3^2 - d^2)}$$

$$\Downarrow$$

$$v_d = \sqrt{36.2572 \cdot (181.3 - d) - 0.0710 \cdot (181.3^2 - d^2)}$$

This equation yields the speed in feet per second. In Table 4.1, the resulting speeds calculated with an average deceleration rate of 0.363g have been converted to both miles per hour and kilometers per hour. Next, Equation (4.28) can be used to calculate the duration associated with each segment and Equation (4.29) can be used to calculate the average roll velocity for each segment. These calculated times and roll velocities are also listed in Table 4.1.

Figure 4.15 is a graph that depicts the calculated roll velocities for both constant deceleration model used in Chapter 3 and for the linearly decreasing deceleration model employed here. The linearly decreasing deceleration rate model is plotted with both average deceleration rates (0.363 and 0.44g). The actual roll velocities from the sensor data are also included on this graph. In this instance, both of the curves that utilized the linearly decreasing deceleration rate model produced a better match with the actual

FIGURE 4.15 Comparison of roll velocity curves calculated with constant and linearly decreasing deceleration rates compared to the actual roll velocities from the sensor data.

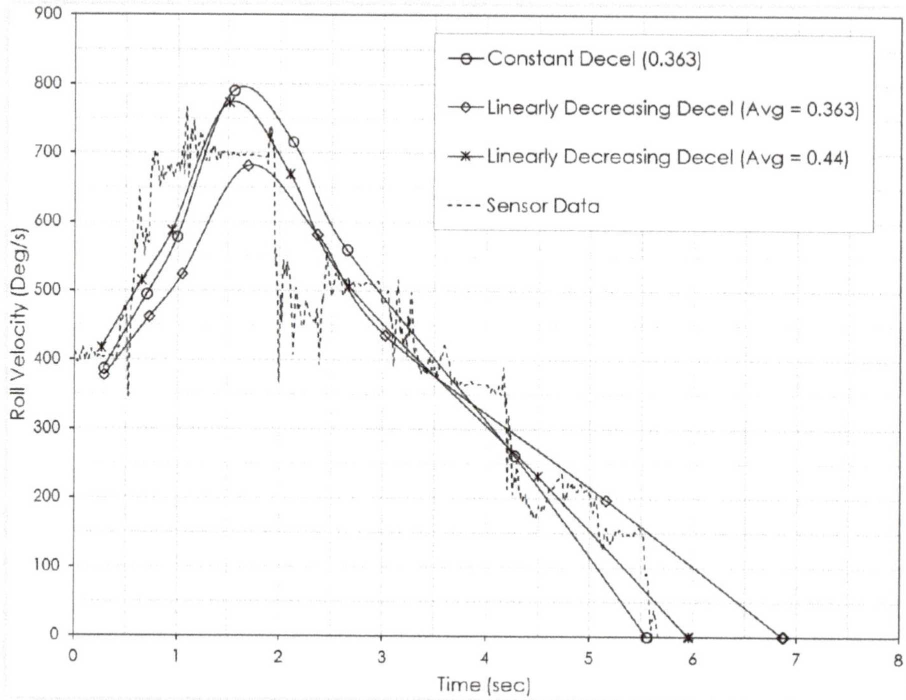

magnitude of the peak roll velocity than did the constant deceleration rate model. The calculations with an average deceleration rate of 0.44g captured the roll duration better than the calculations with an average deceleration rate of 0.363g, but the 0.44g curve overestimated the peak roll velocity more significantly than the 0.363g curve. Figure 4.16 is a similar graph that compares the calculated OTG speeds for the constant deceleration and linearly decreasing deceleration models.

The roll angles listed in Table 4.1 were determined from the physical evidence, but also benefited from review of the video from this test. Thus, these roll angles reflect 7 complete rolls by the vehicle. Again, the question can be posed as to whether analysis would enable the reconstructionist to identify the presence of 7 rolls in this case (without the benefit of video) even though one of those rolls occurred while the vehicle was airborne. Figure 4.17 is a roll velocity graph that contains three curves - the sensor data that gives the measured roll velocities, the reconstructed roll velocity curve assuming a linearly decreasing deceleration rate and 6 rolls, and the reconstructed roll velocity curve assuming 7 rolls. Both of these curves assume an average deceleration rate of 0.363g. The airborne roll occurred during the 4th segment, and so the difference between the reconstructed curves is reflected in the roll velocity calculated for this segment. The reconstructed curve that assumed 6 rolls again exhibits a precipitous drop in roll velocity below its value for the 3rd and 5th segments. The shape of this reconstructed curve does not reflect the characteristics of typical roll velocity curves discussed earlier in this chapter. The reconstructionist could reasonably reject this curve. The reconstructed roll velocity curve that assumed 7 rolls does have the appearance of a typical roll curve. Thus, in this case a reconstructionist could detect the airborne roll even though there would be no physical evidence associated with it.

FIGURE 4.16 Velocity curves calculated with constant and linearly decreasing deceleration rates.

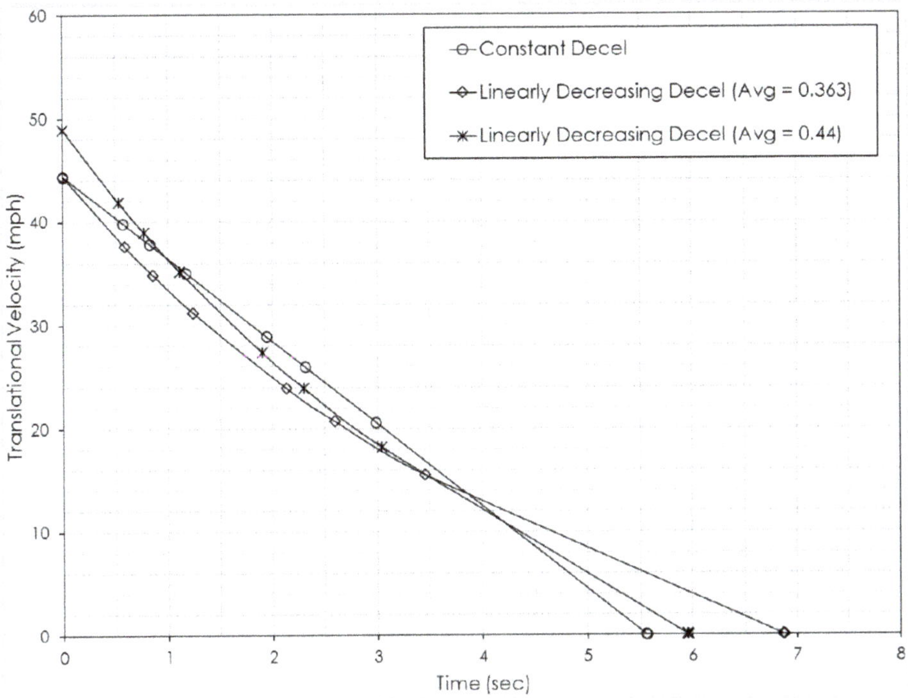

FIGURE 4.17 Comparison of roll velocity curves calculated assuming 6 and 7 rolls compared to the actual roll velocities from the sensor data.

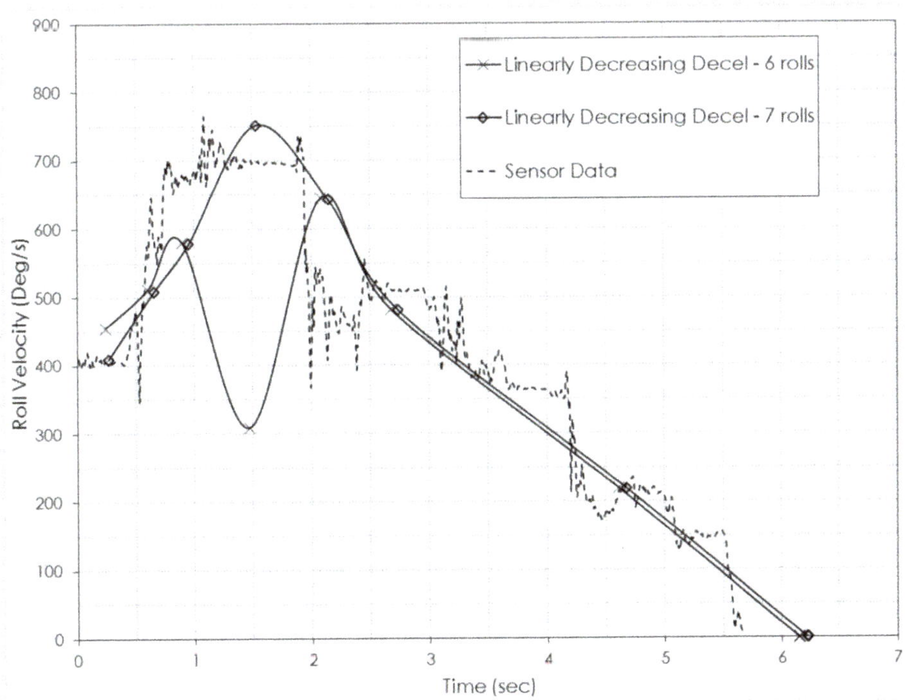

References

1. Rose, N., Beauchamp, G., and Fenton, S., "Factors Influencing Roof-to-Ground Impact Severity: Video Analysis and Analytical Modeling," SAE Technical Paper 2007-01-0726, 2007, doi:10.4271/2007-01-0726.
2. Rose, N., Fenton, S., Beauchamp, G., and McCoy, R., "Analysis of Vehicle-to-Ground Impacts During a Rollover with an Impulse-Momentum Impact Model," *SAE Int. J. Passeng. Cars - Mech. Syst.* 1, no. 1 (2009): 105-123, doi:10.4271/2008-01-0178.
3. Rose, N., Beauchamp, G., and Fenton, S., "The Influence of Vehicle-to-Ground Impact Conditions on Rollover Dynamics and Severity," SAE Technical Paper 2008-01-0194, 2008, doi:10.4271/2008-01-0194.
4. Rose, N., Neale, W., Fenton, S., Hessel, D. et al., "A Method to Quantify Vehicle Dynamics and Deformation for Vehicle Rollover Tests Using Camera-Matching Video Analysis," *SAE Int. J. Passeng. Cars - Mech. Syst.* 1, no. 1 (2009): 301-317, doi:10.4271/2008-01-0350.
5. Rose, N. and Beauchamp, G., "Analysis of a Dolly Rollover with PC-Crash," SAE Technical Paper 2009-01-0822, 2009, doi:10.4271/2009-01-0822.
6. Brach, R., "An Impact Moment Coefficient for Vehicle Collision Analysis," SAE Technical Paper 770014, 1977, doi:10.4271/770014.
7. Fonda, A., "Nonconservation of Momentum during Impact," SAE Technical Paper 950355, 1995, doi:10.4271/950355.
8. Brach, R. and Brach, R., "A Review of Impact Models for Vehicle Collision," SAE Technical Paper 870048, 1987, doi:10.4271/870048.
9. Brach, R., "Energy Loss in Vehicle Collisions," SAE Technical Paper 871993, 1987, doi:10.4271/871993.
10. Brach, R.M. and Brach, R.M., *Vehicle Accident Analysis and Reconstruction Methods*, (Warrendale, PA: SAE, 2005), ISBN:978-0-7680-3437-0.
11. Brach, R.M., *Mechanical Impact Dynamics: Rigid Body Collisions*, (2007), ISBN:978-1-4028-9462-6. This reference is available at www.brachengineering.com.
12. Marine, M., "On the Concept of Inter-Vehicle Friction and Its Application in Automobile Accident Reconstruction," SAE Technical Paper 2007-01-0744, 2007, doi:10.4271/2007-01-0744.
13. Chen, H. and Guenther, D., "Modeling of Rollover Sequences," SAE Technical Paper 931976, 1993, doi:10.4271/931976.
14. Rose, N. and Beauchamp, G., "Development of a Variable Deceleration Rate Approach to Rollover Crash Reconstruction," *SAE Int. J. Passeng. Cars - Mech. Syst.* 2, no. 1 (2009): 308-332, doi:10.4271/2009-01-0093.
15. Carter, J., Luepke, P., Henry, K., Germane, G. et al., "Rollover Dynamics: An Exploration of the Fundamentals," *SAE Int. J. Passeng. Cars - Mech. Syst.* 1, no. 1 (2008): 80-104, doi:10.4271/2008-01-0172.
16. Rose, N.A. and Beauchamp, G., "A Variable Deceleration Rate Approach to Rollover Crash Reconstruction," in *Collision. The International Compendium for Crash Research*, Vol. 1, no. 1, ISSN 1934-8681
17. Luepke, P., Carhart, M., Croteau, J., Morrison, R. et al., "An Evaluation of Laminated Side Window Glass Performance During Rollover," SAE Technical Paper 2007-01-0367, 2007, doi:10.4271/2007-01-0367.

18. Luepke, P., Carter, J., Henry, K., Germane, G. et al., "Rollover Crash Tests on Dirt: An Examination of Rollover Dynamics," *SAE Int. J. Passeng. Cars - Mech. Syst.* 1, no. 1 (2009): 18-30, doi:10.4271/2008-01-0156.

19. Arndt, M., Stevens, D., and Arndt, S., "Comparison of Linear Variable Deceleration Rate Rollover Reconstruction to Steer-Induced Rollover Tests," SAE Technical Paper 2012-01-0469, 2012, doi:10.4271/2012-01-0469.

20. Stevens, D., Arndt, S., Wayne, L., Arndt, M. et al., "Rollover Crash Test Results: Steer-Induced Rollovers," SAE Technical Paper 2011-01-1114, 2011, doi:10.4271/2011-01-1114.

21. Funk, J., "An Integrated Model of Rolling and Sliding in Rollover Crashes," SAE Technical Paper 2012-01-0605, 2012, doi:10.4271/2012-01-0605.

22. Arndt, M., "Drag Factors from Rollover Crash Testing for Crash Reconstructions," *Proceedings of the ASME 2011 International Mechanical Engineering Congress & Exposition IMECE2011*, Denver, CO, November 11-17, 2011, IMECE2011-65537.

23. Asay, A. and Woolley, R., "Rollover Testing of Sport Utility Vehicles (SUVs) on an Actual Highway," SAE Technical Paper 2010-01-0521, 2010, doi:10.4271/2010-01-0521.

Appendix

FIGURE 4.18 Critical impulse ratio for a number of impact scenarios (SI units).

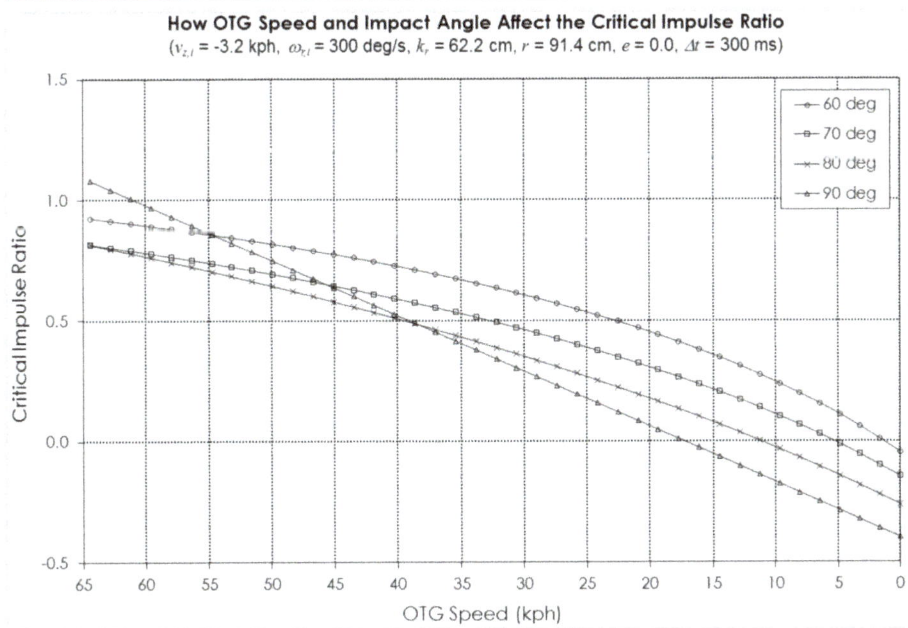

FIGURE 4.19 Impulse ratio for a number of impact scenarios (SI units).

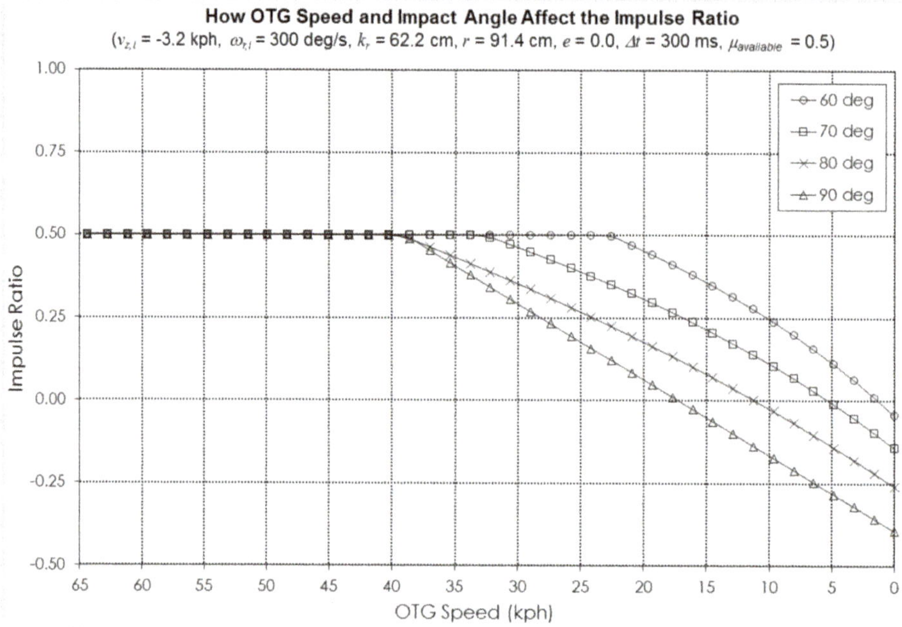

FIGURE 4.20 Comparison of OTG speed obtained with constant deceleration and linearly decreasing deceleration models - Expedition (SI units).

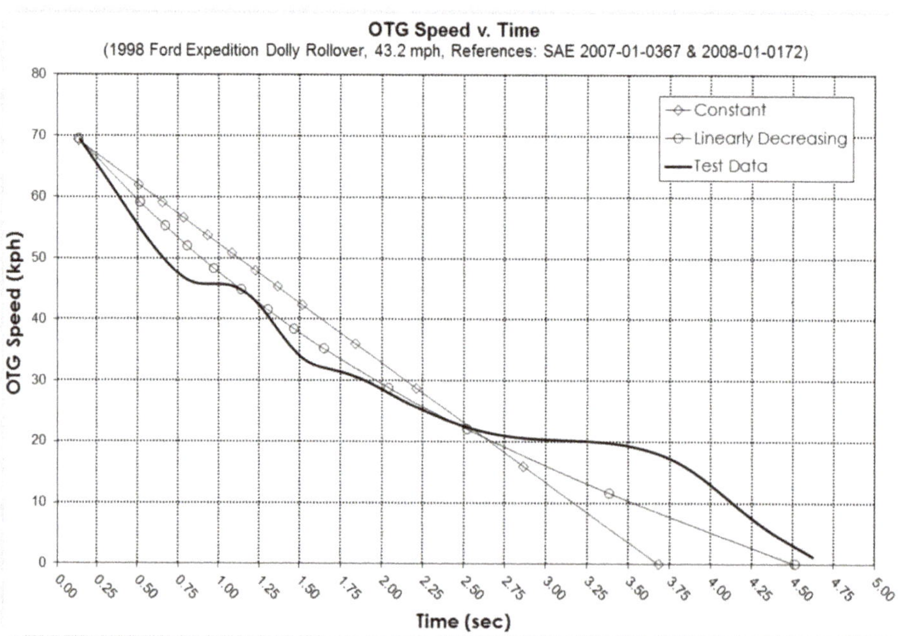

FIGURE 4.21 Comparison of OTG speed obtained with constant deceleration model and linearly decreasing deceleration model - XC90 (SI units).

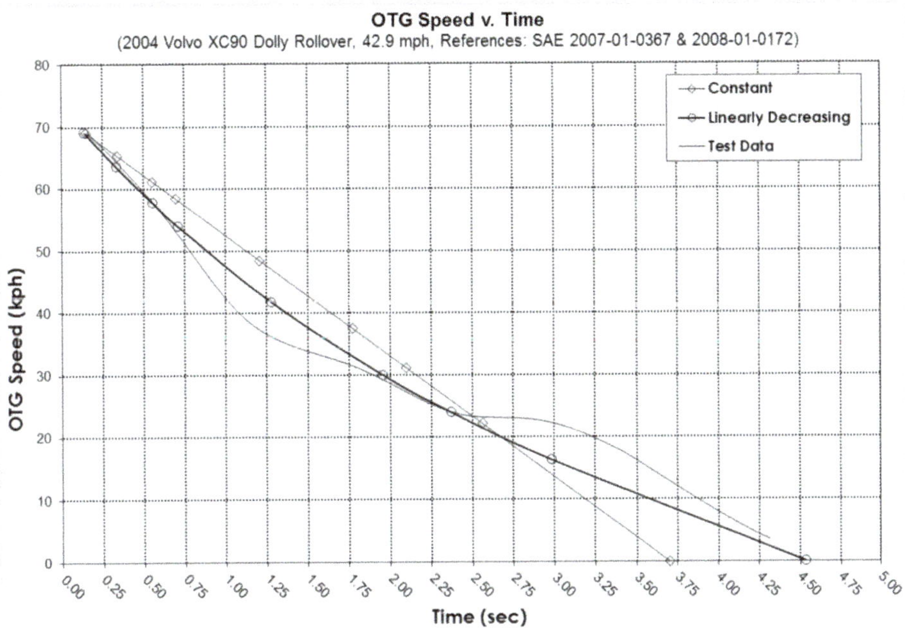

CHAPTER 5

Analysis Methods: Trip Phase

According to Martinez [1], "the trip phase of [a rollover] accident is the portion of the event in which the lateral forces overwhelm the vehicle's ability to maintain four wheeled contact on the ground." During the trip phase, lateral forces applied to the tires or wheels on the leading side of the vehicle cause the trailing side tires to lose contact with the ground and the vehicle to begin tipping up. When the center of mass (CoM) of the vehicle rolls past the point of contact between the leading side tires and the ground, the trip phase ends and the rollover phase begins [2]. During the trip phase, the vehicle acquires an upward velocity and a roll velocity and loses a portion of its ground plane velocity. At the end of the trip phase, the vehicle becomes airborne. Once this occurs, the roll velocity that has been developing during the trip phase levels off and become constant until the next contact with the ground occurs [3].

Following are several examples of mechanisms that can supply the tripping force necessary to initiate a rollover: (1) a combination of tire forces and suspension effects from severe steering inputs [4, 5, 6, 7, 8], (2) interaction between a tire or wheel rim and pavement [9], (3) wheels furrowing into soil or sod [3, 10, 11, 12] and (4) wheels impacting a curb [10, 13, 14]. For each of these mechanisms, the magnitude and duration of the tripping force and the manner in which that force is applied through time varies. The duration of the trip phase and the speed changes that occur during this phase depend on the characteristics of the trip force [2]. To the extent that the characteristics of the tripping force affect the characteristics of the vehicle motion during the trip phase, each tripping mechanism produces trip phase vehicle motion characteristic of that mechanism. The features of the vehicle motion during the trip phase also depend on the vehicle characteristics, such as the weight, moments of inertia, CoM height, track width, and suspension characteristics. This chapter develops methods for calculating the tripping forces for a particular rollover and for determining the changes in over-the-ground, vertical, and roll speeds that occur during the trip phase.

Trip Modeling and Data

Cooperrider examined the trip phase both experimentally and analytically [10]. He conducted a series of rollover tests that included both curb-tripped and soil-tripped rollovers. Two out of five curb trip tests resulted in rollover. Both soil trip tests resulted in rollover. The curb trip tests utilized a 6-in. (15 cm) square section of steel box tubing that was rigidly attached to the ground perpendicular to the vehicle's travel direction. The test vehicles traveled laterally into the curb with their passenger's (right) sides leading at speeds around 30 mph (48 kph). In the two tests where rollover occurred as a result of the curb impact, the vehicles experienced average deceleration rates during the trip of 12.4g and 13.2g. These decelerations occurred over time periods of 60 and 40 ms. The soil-tripped tests utilized a 40 ft by 20 ft (12.2 by 6.1 m) area of loosely packed dirt and finely crushed granite. The test vehicles traveled laterally into this soil with their right sides leading. The initial speeds of the vehicles in these two tests were 33.7 and 27.0 mph (54.2 and 43.5 kph). The vehicle tires furrowed into the soil, causing the vehicles to rollover. In these two tests, the vehicles experienced average deceleration rates during the trips of 1.62g and 1.71g. These decelerations occurred over time periods of 513 and 460 ms.

Cooperrider also derived an expression for the lateral force necessary to initiate and sustain the trip phase. This expression was derived by first assuming that the tripping force is constant and then writing an approximate set of equations of motion for a model similar to that depicted in Figure 5.1. The vehicle depicted in this figure is in the midst of the trip phase, with the trailing wheels having lifted off the ground and with the lateral tripping force applied to the leading side tires. This model assumes a rigid vehicle, and so, suspension and tire compliances are ignored. The model depicted in Figure 5.1 is only different from the one Cooperrider used in the orientation of the coordinate system. Cooperrider had the y direction oriented to the right and the z direction oriented down (standard SAE coordinate system). The model below reverses both of these, using a right-handed coordinate system (RHS) with the y-axis pointed to the left and the z-axis pointed up. The angular position of the body is measured with positive rotation in the clockwise direction. With this RHS coordinate system, the vehicle depicted in Figure 5.1 would be facing into the page and rolling with its passenger's (right) side leading.

FIGURE 5.1 Free-body diagram for a rigid vehicle during the trip phase.

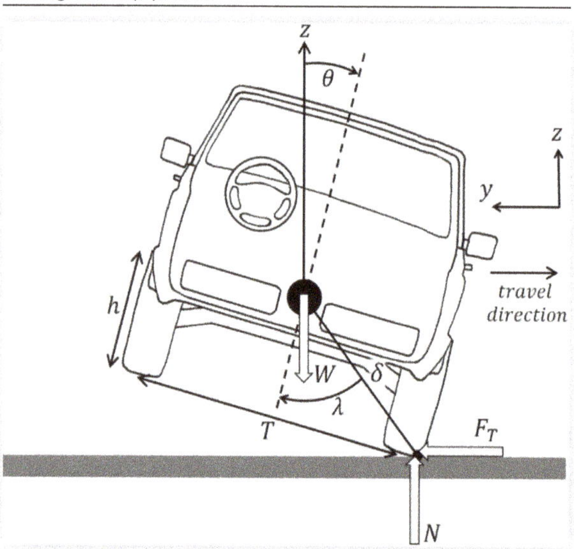

The equations of motion that Cooperrider wrote for this model are as follows (using the coordinate system depicted in Figure 5.1):

$$m\ddot{y} = F_T \tag{5.1}$$

$$m\ddot{z} = N - mg \tag{5.2}$$

$$I_{xx}\ddot{\theta} = F_T \cdot h - N \cdot \frac{T}{2} \tag{5.3}$$

In these equations,

- y is the lateral position of the vehicle's CoM
- z is the vertical position
- θ is the vehicle's roll angle
- F_T is the tripping force
- N is the normal force
- h is the height of the vehicle CoM
- T is the vehicle track width
- I_{xx} is the roll moment of inertia

Equation (5.3) is approximate because the moment arms employed are only accurate when the roll angle is zero. This assumption is only true at the instant the trailing side wheels lift off. These moment arms become increasingly inaccurate as the roll angle increases. Thus, Cooperrider's model is set up to predict the lateral force necessary to initiate trailing wheel liftoff.

Cooperrider next wrote the following geometric constraint equation which imposes the condition that the leading side tires remain in contact with the ground during the trip phase:

$$z = \frac{T}{2} \cdot \sin\theta \tag{5.4}$$

Cooperrider then employed a small angle assumption for Equation (5.4) reducing it to the following equation. This equation also only holds true at the instant the trailing side tires lift off and, from that point forward, it becomes increasingly inaccurate:

$$z = \frac{T}{2}\theta \tag{5.5}$$

Cooperrider used these equations in conjunction with the idea that in order to initiate a rollover, "the kinetic energy in the roll and vertical motions must equal the remaining change in potential energy needed to raise the [center of mass] to the neutral stability point...." Cooperrider reported the following expression that relates the trip force and the trip duration (Δt):

$$\left(\frac{F_T}{mg} - \frac{T}{2h}\right)^2 \cdot \Delta t^2 + \frac{T}{2h} \cdot \left(\frac{F_T}{mg} - \frac{T}{2h}\right) \cdot \Delta t^2 = \frac{2\bar{I}_{xx}}{mgh} \cdot \left(\sqrt{\left(\frac{T}{2h}\right)^2 + 1} - 1\right) \tag{5.6}$$

In this equation,

$$\bar{I}_{xx} = I_{xx} + \frac{mT^4}{4} \tag{5.7}$$

Since a reconstructionist is typically interested in the magnitude of the tripping force, which is directly related to the vehicle's deceleration rate, this expression is not in an easily useable form. Nonetheless, Cooperrider reported that "the correlation of the four [curb and soil trip] data points provided by [the experimental data] with the constant force model is good, although additional comparisons are needed to conclusively evaluate the model."

Brach examined options for modeling the trip phase in his accident reconstruction text [15]. As a part of his discussion, he presented a reworked form of Cooperrider's equation, writing it in a more useable form that explicitly yields the tripping force in gravitational units. He also utilized the roll radius of gyration for the vehicle, noting that $I_{xx} = mk_{xx}^2$. Brach's equation is as follows:

$$\frac{F_T}{mg} = \frac{1}{2\Delta t}\left[\Delta t \frac{T}{2h} + \sqrt{\Delta t^2 \left(\frac{T}{2h}\right)^2 + \frac{8h}{g}\left(\frac{k_{xx}^2}{h^2} + \left(\frac{T}{2h}\right)^2\right)\left(\sqrt{\left(\frac{T}{2h}\right)^2 + 1} - 1\right)}\right] \tag{5.8}$$

Equation (5.8) was used to calculate lateral tripping forces for the two vehicles used in Cooperrider's 1990 curb- and soil-tripped tests. In the two soil-tripped tests, Cooperrider reported that the vehicles experienced average deceleration rates during the trips of 1.62g and 1.71g. These decelerations occurred over time periods of 513 and 460 ms. Equation (5.8) yielded trip forces of 1.84g and 1.82g, higher than the actual values. These calculated values yield speed changes of 20.7 and 18.3 mph, whereas the actual reported decelerations would yield speed changes of 18.2 and 17.3 mph. In the two curb-tripped tests where rollover occurred, with the same vehicle models as in the soil-tripped tests, the vehicles experienced average deceleration rates during the trip of 12.4g and 13.2g. These decelerations occurred over time periods of 60 and 40 ms. For these durations, Equation (5.8) yielded trip forces or deceleration rates of 8.89g and 11.93g, lower than actual. These calculated values would yield speed changes of 11.7 and 10.5 mph, whereas the actual reported decelerations would yield speed changes of 16.3 and 11.6 mph.

In 1998, Cooperrider reported six additional soil trip tests that each utilized a 1984 Oldsmobile Cutlass Ciera. The first test was run with an initial speed of 13.5 mph (21.7 kph) and, as Cooperrider stated, "the subsequent test speeds were established based on the previous test results in an effort to closely bracket the minimum trip speed." Ultimately, the initial speeds for the vehicles in this test series ranged from 13.5 to 42.9 mph (21.7 to 69.0 kph). In the first four tests, the vehicle did not roll over. The initial speed of the test vehicle in the fourth test was 21.2 mph (34.1 kph). In the fifth and sixth tests, which had speeds of 23.0 and 42.9 mph (37.0 and 69.0 kph), the vehicle rolled over - ½ roll in the fifth test and 1 roll in the sixth test. When the trip phase was defined as the time over which the vehicle reached a roll velocity of 100°/s, the average deceleration rate of the vehicles during the trip phase of both tests where rollover occurred was 1.23g (durations of 624 and 664 ms). When the trip phase was defined as the time over which the vehicle reached a roll angle of 52°, the average deceleration rates from the trip phase of the two tests were 1.05g and 1.39g (durations of 784 and 792 ms). Neither of these definitions of the trip phase could be employed in practice.

Equation (5.8) was used to calculate lateral tripping forces for the two tests that resulted in rollover in Cooperrider's 1998 article. When the trip phase was defined as the time over which the vehicle reached a roll angle of 52°, the trip phase durations were 784 and 792 ms. For these trip durations, Equation (5.8) yielded trip deceleration rates of 1.57g and 1.56g, higher than the values of average deceleration rates Cooperrider reported. These calculated deceleration rates resulted in calculated speed changes of 26.9 and 27.1 mph, whereas the reported speed changes were 18.1 and 24.2 mph. Thus, for the four soil-tripped tests reported by Cooperrider, his model overestimated the deceleration rate, and for the two curb trip tests, it underestimated the deceleration rate.

Larson [4] reported two soil-tripped rollover tests - one with a 1991 Jeep Wrangler and another with a 1986 Chevrolet S-10 Blazer. Each vehicle was released without an initial roll angle and a slip angle of approximately 52° across a dirt surface with a 15.5° downslope. The vehicles deposited tire and rim furrows in the dirt before beginning to roll over. Larson reported that the Jeep Wrangler was traveling 40.7 mph (65.5 kph) when it was released from the dolly, that it traveled 70.5 ft (21.5 m) from release to rest, that it furrowed for approximately 20.7 ft (6.3 m) before entering the roll phase, and that it completed one roll before coming to rest. The roll distance was approximately 49.8 ft (15.2 m). Larson did not report a speed for the vehicle at the end of the furrowing. The average deceleration rate for the trip and roll phases combined was 0.78g. If we assume that the deceleration rate of this vehicle during the roll phase was around 0.45g, then the deceleration rate during the trip phase was around 1.59g.

Larson reported that the Chevrolet S-10 Blazer was traveling 41.1 mph (66.1 kph) when it was released from the dolly, that it traveled 74 ft (22.6 m) from release to rest, that it furrowed for approximately 18.1 ft (5.5 m) before entering the roll phase, and that it completed two rolls before coming to rest. The roll distance was approximately 55.9 ft (17.1 m). Again, Larson did not report a speed for the vehicle at the end of furrowing. The average deceleration rate for the trip and roll phases combined was 0.76g. If we assume that the deceleration rate of this vehicle during the roll phase was around 0.45g, then the deceleration rate during the trip phase was around 1.73g.

Equation (5.8) was used to analyze Larson's soil trip test of a Chevrolet S-10 Blazer based on the reported furrow distance and initial speed. This equation yielded a lateral tripping force of 1.64g. Here, though, the vehicle had a slip angle of approximately 52° during the trip phase. Thus, only a portion of the lateral tripping force would have been opposing the velocity and acting to decelerate the vehicle. Equation (5.9) can be used to take this into account:

$$f_{trip} = f_{cooperrider} \cdot \sin\alpha + f_{long} \cdot \cos\alpha \tag{5.9}$$

In this equation

f_{trip} is the resultant deceleration during the trip phase
$f_{cooperrider}$ is the normalized lateral tripping force calculated with Equation (5.8)
α is the average vehicle slip angle during the trip phase
f_{long} is any longitudinal force (normalized by the vehicle weight) that was applied to the vehicle during the trip phase from braking or furrowing effects

Using this equation for Larson's test with the assumption that there was no longitudinal force results in a deceleration rate of 1.29g, significantly below the actual value of 1.73g.

In 2002, Hughes reported a series of 8 curb-tripped rollover tests conducted with 2001 Nissan Pathfinders. These tests used a modified dolly fixture that did not introduce an initial roll angle to the test vehicles. Each test vehicle was launched laterally from the dolly with a speed of around 31 mph (50 kph) and then the leading side tires impacted the curb, which was 6 in. (15 cm) high. Rollover occurred in all 8 tests. The roll distances ranged from 54 to 72 ft (16 to 22 m), and the number of rolls varied between 1 and 2½ times. Peak accelerations during the curb impacts ranged from 14.5 to 20.5g. Hughes did not report impact durations.

Asay reported some analysis of decelerations experienced while in the trip phase for each of the tests he published in 2010. He observed that, for all of the tests where rollover ultimately occurred, "the tripping phase was well defined by distinct furrows in the soil that terminated as the vehicle overturned." He reported that the peak lateral acceleration experienced by the vehicles during the trip phase in these tests ranged from 2.76 to 5.70g. The peak lateral accelerations while furrowing through the dirt for the tests that did not ultimately result in rollover were 1.76 and 1.71g.

Trip Modeling without Small Angle Assumptions

Rose and Beauchamp [16] reported modeling of the trip phase using the model depicted in Figure 5.1. Their modeling did not employ small angle assumptions, and so, it can be used to determine the effects of Cooperrider's use of small angle assumptions. The external forces applied to the vehicle in this model again include the weight, the lateral tripping force, and the leading-side normal force. The vehicle weight is applied at the CoM, and the lateral tripping force and the normal force are applied at the leading side wheels. This has the following equations of motion:

$$ma_y = F_T \tag{5.10}$$

$$ma_z = N - mg \tag{5.11}$$

$$I_{xx}\alpha_r = F_T \cdot \delta \cdot c(\lambda - \theta_r) - N \cdot \delta \cdot c(\lambda - \theta_r) \tag{5.12}$$

In Equations (5.10) through (5.12)

a_y is the CoM ground plane acceleration
a_z is the CoM vertical acceleration
α_r is the roll axis angular acceleration of the vehicle
m is the vehicle mass
I_{xx} is the vehicle roll moment of inertia
F_T is the time dependent tripping force
N is the normal load at the leading wheels
θ_r is the vehicle roll angle
δ is a geometric parameter defined as follows:

$$\delta = \sqrt{h^2 + \left(\frac{T}{2}\right)^2} \tag{5.13}$$

Thus, δ is the distance from the vehicle CoM to the point of the application of the trip force. The s and c in Equations (5.10) through (5.12) refer to the sine and cosine. The angle λ in Equation (5.12) is defined as follows:

$$\lambda = \tan^{-1}\left(\frac{T}{2h}\right) \tag{5.14}$$

During the trip phase, the leading side wheels remain in contact with the ground, and thus, Equations (5.10) through (5.14) are supplemented with the following geometric constraint equation:

$$z = \delta \cdot c(\lambda - \theta_r) \tag{5.15}$$

Differentiation of Equation (5.15) twice with respect to time yields Equations (5.16) and (5.17), which give the vertical velocity and vertical acceleration of the CoM, respectively. The symbol ω_r represents the roll velocity:

$$v_z = \delta \cdot \omega_r \cdot s(\lambda - \theta_r) \tag{5.16}$$

$$a_z = \delta \cdot \alpha_r \cdot s(\lambda - \theta_r) - \delta \cdot \omega_r^2 \cdot c(\lambda - \theta_r) \tag{5.17}$$

Substitution of Equation (5.17) into Equation (5.11) yields Equation (5.18), which describes the leading-side normal load as a function of the vehicle weight, roll angle, roll velocity, and roll acceleration:

$$N = mg + m\delta\alpha_r \cdot s(\lambda - \theta_r) - m\delta\omega_r^2 \cdot c(\lambda - \theta_r) \tag{5.18}$$

Substitution of Equation (5.18) into Equation (5.12) leads to Equation (5.19), a nonlinear, second-order ordinary differential equation describing the angular motion of the rigid vehicle:

$$\alpha_r = \frac{F_T\delta c(\lambda - \theta_r) - mg\delta s(\lambda - \theta_r) + m\delta^2\omega_r^2 s(\lambda - \theta_r)c(\lambda - \theta_r)}{I_{xx} + \delta^2 s^2(\lambda - \theta_r)} \tag{5.19}$$

Numerical integration of Equation (5.19) yields a solution for any particular case. In this process, the trip duration is prescribed, and the magnitude of the tripping force is iteratively changed until the normal load on the leading tires goes to zero at the prescribed trip duration. This analysis would usually begin with the initial conditions of a zero roll angle and zero roll velocity.

Figures 5.2 and 5.3 show analysis conducted with both the Cooperrider model and the Rose and Beauchamp model. This analysis assumed a constant trip force throughout

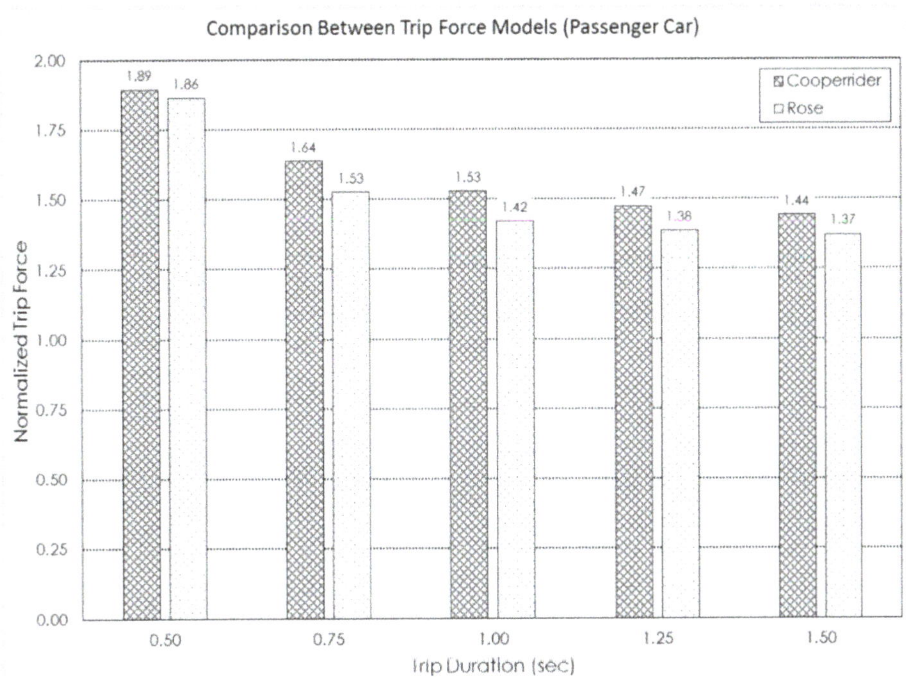

FIGURE 5.2 Comparison between trip models with and without small angle assumptions (passenger car parameters).

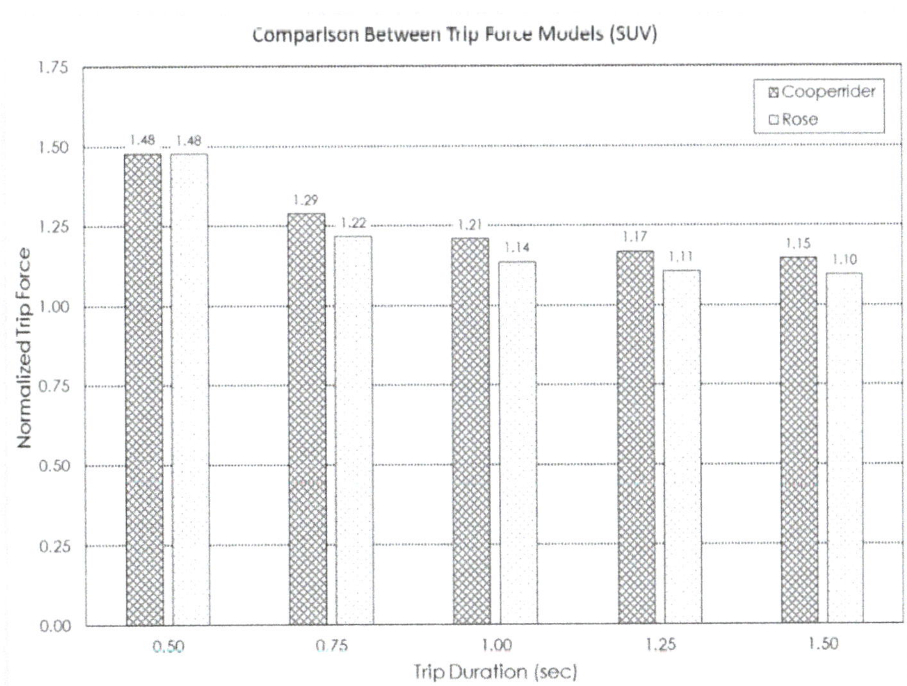

FIGURE 5.3 Comparison between trip models with and without small angle assumptions (SUV parameters).

CHAPTER 5

the trip phase. Figure 5.2 compares the normalized trip force obtained with the two models for parameters typical of a passenger vehicle. Figure 5.3 contains a similar comparison for parameters typical of an SUV. These results were developed using the following vehicle parameter sets:

- **SUV Parameter Set**
  - Vehicle Weight = 5000 lb (2268 kg)
  - Roll Moment of Inertia = 750 lb-ft-s2 (104 kg-m-s2)
  - Radius of Gyration = 2.2 ft (.67 m)
  - Track Width = 61.0 in. (155 cm)
  - CoM Height = 28.0 in. (71.1 cm)
- **Passenger Car Parameter Set**
  - Vehicle Weight = 3600 lb (1633 kg)
  - Roll Moment of Inertia = 500 lb-ft-s2 (69.1 kg-m-s2)
  - Radius of Gyration = 2.1 ft (.64 m)
  - Track Width = 60.0 in. (152 cm)
  - CoM Height = 22.0 in. (55.9 cm)

In these figures, the trip duration was varied between 0.5 and 1.5 s, in ¼ s increments. The values depicted in these graphs show that the small angle assumptions employed by Cooperrider result in higher calculated trip forces than are obtained with a model that does not employ small angle assumptions. That said, the differences are small. The Cooperrider model - in the reworked form presented by Brach - is easier to implement than a model that does not employ small angle assumptions, so given the small differences between the models, it makes sense for accident reconstructionists to use the Cooperrider model for trip phase speed calculations, assuming that the rigid vehicle assumptions do not make it inaccurate. This possibility is explored later.

Rose and Beauchamp also conducted analysis with both a linearly increasing trip force and with a sinusoid-shaped trip force. Figure 5.4 contains roll angle curves, Figure 5.5 contains roll velocity curves, and Figure 5.6 contains vertical velocity curves, obtained with both a tripping force that increases linearly with time, as described by Equation (5.20), and with a sinusoidal tripping force described by

FIGURE 5.4 Roll angle versus time, linearly increasing trip force (left) and sinusoidal trip force (right).

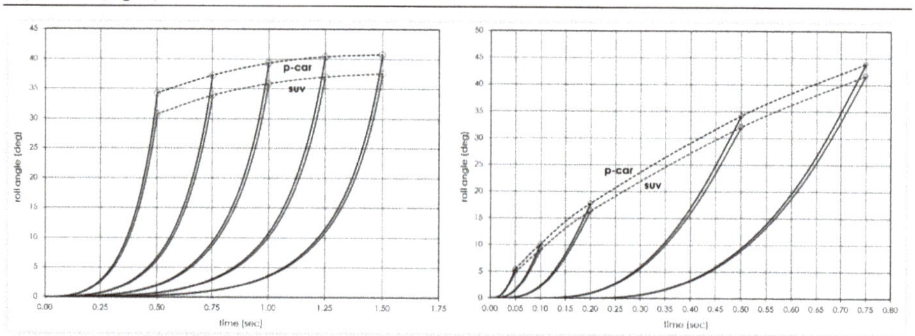

FIGURE 5.5 Roll rate versus time, linearly increasing trip force (left), and sinusoidal trip force (right).

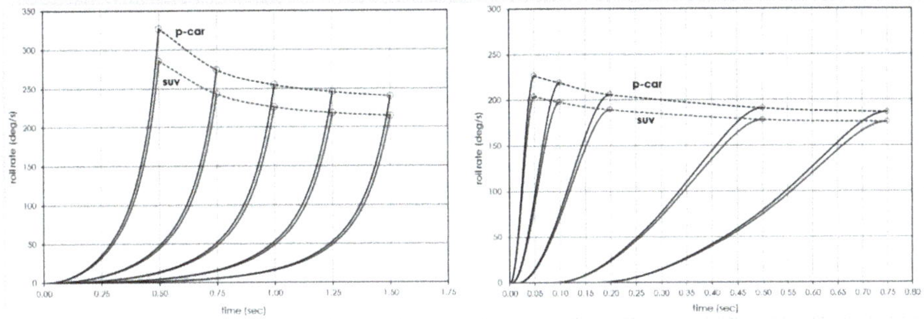

FIGURE 5.6 Vertical velocity versus time, linearly increasing trip force (left) and sinusoidal trip force (right).

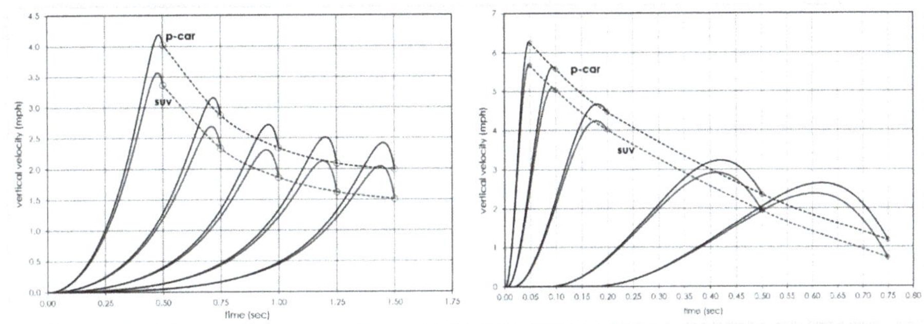

Equation (5.21). A version of Figure 5.6 containing SI units is included in the appendix to this chapter:

$$F_T(t) = F_0 + (F_{max} - F_0) \cdot \frac{t}{\Delta t_{trip}} \tag{5.20}$$

$$F_T(t) = F_{max} \cdot \sin\left[\pi \frac{t}{\Delta t_{trip}}\right] \tag{5.21}$$

In these equations

F_0 is the initial tripping force magnitude

F_{max} is the maximum tripping force magnitude

Δt_{trip} is the trip duration

t is time

A linearly increasing trip force characteristic could potentially correspond to that which would be produced by a vehicle furrowing into soil, with the force increasing as the depth of the furrowing increases [2, 17]. A sinusoidal trip force shape could potentially represent an impact-type tripping mechanism where the tripping force starts out at zero, builds up to a peak value, and then returns to zero. For each curve in Figures 5.4 through 5.6, the initial tripping force magnitude was set at the value of the static stability factor and the maximum tripping force magnitude was changed until leading wheel liftoff occurred at the prescribed trip duration.

For the linearly increasing trip force, the trip duration was varied between 0.5 and 1.5 s, in ¼ s increments. For the sinusoidal tripping force, the trip duration was varied between 0.05 and 0.75 s. This range starts and ends at lower trip durations than those used for the linearly increasing trip force. The motivation for the low end of this range is the fact that the sinusoidal-type tripping force would likely be associated with impact-type tripping mechanisms, such as curbs, and would generally occur with short-duration applications [10, 13, 18]. The high end of this range was set so that there would be some overlap between the durations used for the linearly increasing and sinusoid-shaped trip force shapes.

Figures 5.4 through 5.6 lead to the following observations:

- An increase in trip duration results in an increase in roll angle and a decrease in the roll and vertical velocities at the end of trip.
- The linearly increasing trip force resulted in higher roll rates for a given trip duration than the sinusoidal trip force. This could mean that soil-tripped rollovers would generally begin with higher roll rates than soil-tripped rollovers. That said, soil-tripped and curb-tripped rollovers are associated with very different trip durations, so the differences in roll rate for the two mechanisms may not be easily discernable in the real world. Ultimately, the physical evidence and the characteristics of a particular vehicle and trip mechanism should drive the reconstructionists' determination of the roll rate with which the vehicle begins the roll phase.
- The sinusoidal trip force resulted in higher vertical velocities than the linearly increasing trip force. This implies that curb-tripped rollovers begin with greater vertical velocity than soil-tripped rollovers. The vertical velocity curves peaked prior to the end of the trip phase. The vertical velocity then dropped off before leading wheel liftoff occurred.
- For a sinusoidal trip force with a short-duration trip phase, leading wheel liftoff occurs at low roll angles relative to what occurred for the longer linearly increasing trip forces. This likely means that curb-tripped rollovers would begin with a lower initial roll angle than soil-tripped rollovers. This finding is consistent with the roll angles observed at the end of the trip in the curb-tripped rollovers reported by Hughes [14]. In the eight tests that Hughes ran, these roll angles ranged from 5 to 12°.

These results demonstrate that the vehicle orientation and velocity conditions that exist when a vehicle enters the roll phase depend on both the vehicle characteristics and the trip force characteristics. It stands to reason that the vehicle motion that occurs during the roll phase depends on the orientation and velocity conditions with which the vehicle enters that phase. Therefore, the trip force characteristics are influential in determining the vehicle motion that occurs during the roll phase. This is not to say that the analyst will be able to discern or define the relationship between the initial conditions for the roll phase and the roll motion that results.

Critical Sliding Velocity

Occasionally, an accident reconstructionist will need to evaluate the speed necessary to cause a curb-tripped rollover. A number of authors have developed models to calculate the critical sliding velocity, which is the lateral velocity at which a vehicle can be tripped by a sideways impact between the vehicle wheels and a fixed obstacle, such as a curb, and roll ¼ of a revolution. For example, Jones [19] presented the following equation to calculate the critical sliding velocity:

$$CSV = \sqrt{\frac{2gI_0}{mh}\left(\sqrt{1+\left(\frac{T}{2h}\right)^2}-1\right)} \tag{5.22}$$

In this equation

- g is the gravitational constant
- m is the vehicle mass
- I_0 is the roll moment of inertia about the point at which the force is applied to the wheels of the vehicle
- h is the center of gravity height
- T is the track width

The roll moment of inertia about the point of the force application is given by the following equation:

$$I_0 = I_{cg} + m\left(\left(\frac{T}{2}\right)^2 + h^2\right) \tag{5.23}$$

In this equation, I_{cg} is the roll moment of inertia about the center of gravity. Using an approximation for the roll moment of inertia about the center of gravity, Lund and Bernard [20] developed the following equation for estimating the critical sliding velocity:

$$CSV = \sqrt{\frac{8.43gh}{3}\left(1+\left(\frac{T}{2h}\right)^2\right)\left(\sqrt{1+\left(\frac{T}{2h}\right)^2}-1\right)} \tag{5.24}$$

These equations assume that the wheels do not break off their hubs. If this occurs, the force generated may be insufficient to cause the vehicle to rollover, even at the speeds predicted by these equations.

Case Study: Asay's Isuzu Rodeo Rollover Test (Test #2, 2010-01-0521)

This section analyzes the trip phase of the Isuzu Rodeo rollover test that Asay reported in 2010. This data allowed for the natural uninterrupted transition between the phases of the rollover to be analyzed. The roll phase was analyzed in Chapters 3 and 4. In this test, the Isuzu was initially traveling a speed of 73.5 mph (118.3 kph). The vehicle was steered sharply to the left, causing the vehicle to travel across the roadway to the left and to yaw counterclockwise. The vehicle developed a slip angle of sufficient magnitude that tire marks were deposited on the road surface. Approximately 1 s later, the first steering input was followed by a significant steering input back to the right. The vehicle reversed its yaw direction but continued off the left side of the road into the dirt, developing a significant slip angle as it yawed in a clockwise manner. The vehicle deposited furrows in the dirt and then began rolling over. Figure 5.7 contains frames from the video of this test that depict the final yaw motion of the vehicle that preceded the roll. The trip phase encompasses all or part of this final yaw.

FIGURE 5.7 Frames from the video of the trip phase.

Based on Asay's survey data, the tire marks and furrows deposited by the Isuzu during the loss-of-control and trip phases were diagrammed, and the vehicle positions depicted in Figure 5.8 were determined from that evidence. The trip phase occurs during the final yaw of the vehicle leading into the roll phase. This final yaw motion is depicted with 5 numbered vehicle positions in Figure 5.8. To determine the actual speed at each of these positions, the VBOX speed data was aligned to these positions in accordance with the alignment data reported by Asay. For the first four positions of the final yaw, the vehicle had not developed significant roll velocity, and so, correction of the speeds for these positions was not necessary. However the fifth position had more roll velocity, so Arndt's corrected speed of 44.4 mph was used for that position. The process of aligning the VBOX data resulted in the following speeds for the five positions identified in Figure 5.8 -68.9 mph at the first position, 67.6 mph at the second position, 64.3 mph at the third position, and 61.1 mph at the fourth position.

FIGURE 5.8 Loss-of-control and trip phase positions determined from the scene evidence.

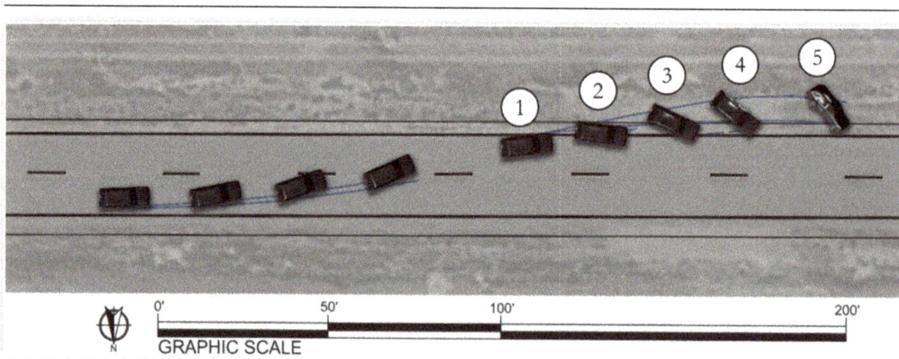

FIGURE 5.9 Measurements for loss-of-control and trip phase positions.

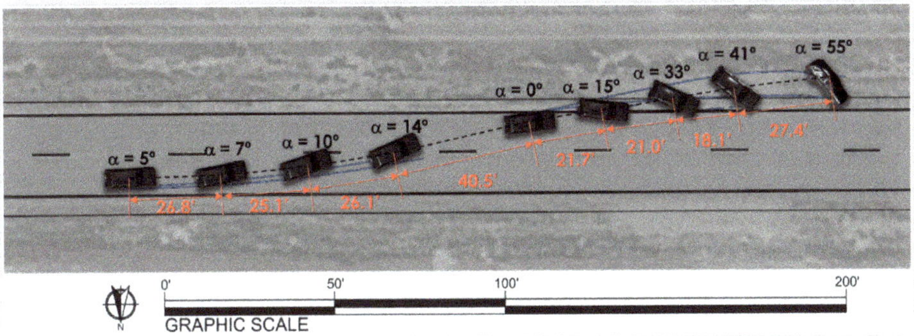

One issue that arises in modeling the trip phase is deciding what portion of the final yaw constitutes the trip phase. One approach would be to look at the tire mark evidence to determine when both trailing side tires have lifted off. In this case, that occurred at Position 4. Another option would be to use the position at which the first trailing side tire lifts off as the beginning of the trip phase. In this case, this occurs at the passenger side rear wheel around Position 3 in Figure 5.8. Still another option would be to treat the entire final yaw from Positions 1 through 5 as the trip phase. The calculations reported in this section considered each of these possible definitions of the trip phase.

The segment distance and slip angle measurements used in these calculations are depicted in Figure 5.9, and the first set of calculations are summarized in Table 5.1. These calculations used Equations (5.8) and (5.9), assumed a deceleration rate during the roll phase of 0.363 to match the speed at the beginning of the roll phase, and used the following values for the Isuzu Rodeo:

- Weight = 4,266 lb
- Roll Moment of Inertia = 518 lb-ft-s^2
- Radius of Gyration = 1.98 ft
- Track Width = 57.5 in.
- CoM Height = 25.9 in.

The calculations summarized in Table 5.1 were initially run assuming there was no longitudinal tripping force present during the trip phase. Then, the longitudinal tripping force that would be necessary to match the actual deceleration rate during the trip phase was calculated. Several observations can be made based on the results reported in Table 5.1. First, with the assumption that there was no longitudinal force present during the trip phase, Equations (5.8) and (5.9) underestimated the actual deceleration rate regardless of how the trip phase was defined. There was no braking applied to the vehicle during this test, so whatever longitudinal forces were present must have been generated due to effects of furrowing and steering. Including these forces would improve the accuracy of the speed analysis.

TABLE 5.1 Summary of trip phase calculations (assumes roll phase deceleration rate of 0.363g)

Trip phase	Actual speed at beginning of trip (mph)	Trip distance (ft)	Actual deceleration rate (g)	Calculated normalized lateral tripping force (g)	Average slip angle (°)	Calculated deceleration rate assuming no longitudinal force	Calculated speed assuming no longitudinal force (mph)	Implied longitudinal deceleration rate (g)
Positions 4-5	61.1	27.4	2.14	1.68	48	1.23	54.6 (−6.5)	1.35
Positions 3-5	64.3	45.5	1.58	1.38	44	0.93	57.0 (−7.3)	0.88
Positions 1-5	68.9	88.2	1.05	1.20	28	0.57	58.9 (−10.0)	0.57

Figure 5.10 shows the actual longitudinal and lateral (ground plane) accelerations measured during the initial counterclockwise yaw of the test vehicle and the subsequent clockwise yaw that includes the trip phase. This graph begins at a time of 0.5 s and continues until a time of 3.25 s. According to Asay's article, time zero corresponds to the release of the vehicle. The ground plane lateral accelerations shown in this graph were calculated from the measured vehicle frame lateral and vertical accelerations and the roll angle. The extents of the final yaw in Figure 5.10 can be identified based on the roll velocity data from the test, the location of tire marks and furrows from the final yaw, and review of the video.

Figure 5.11 shows the relevant portions of the roll velocity data. As this graph shows, the roll velocity builds up during the trip phase over an approximately ½ s time frame

FIGURE 5.10 Longitudinal and lateral acceleration during trip (corrected for roll).

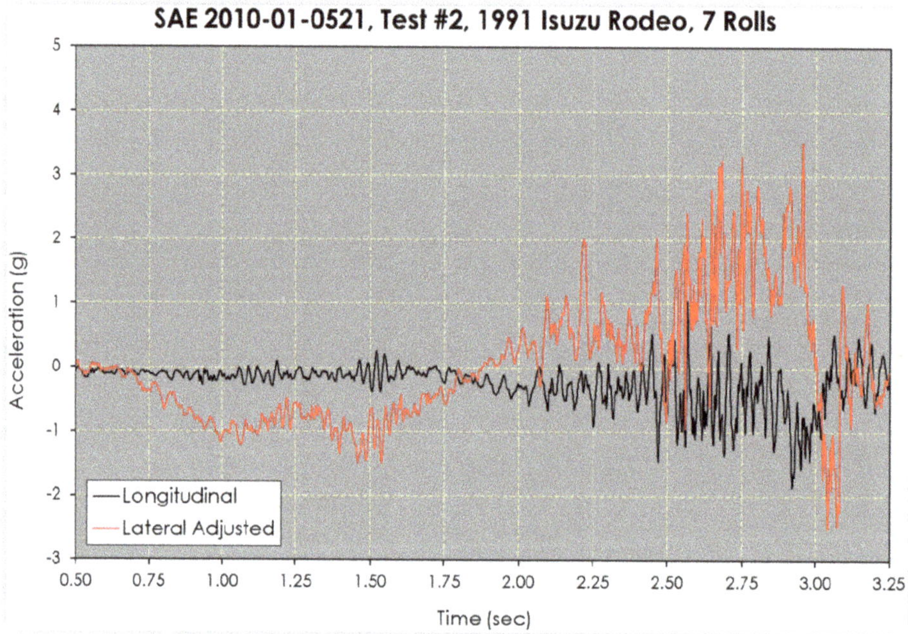

FIGURE 5.11 Roll velocity data.

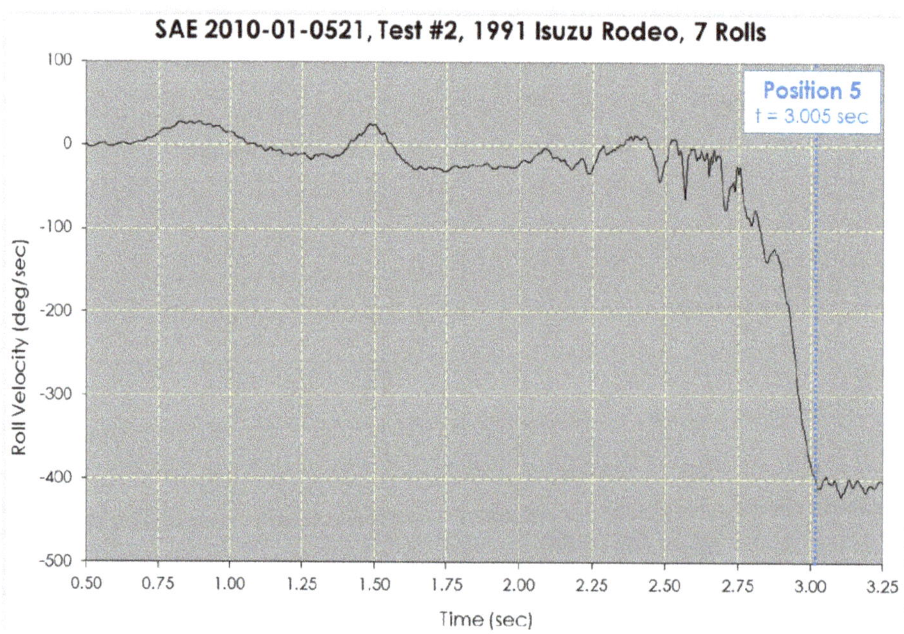

FIGURE 5.12 Roll angle data.

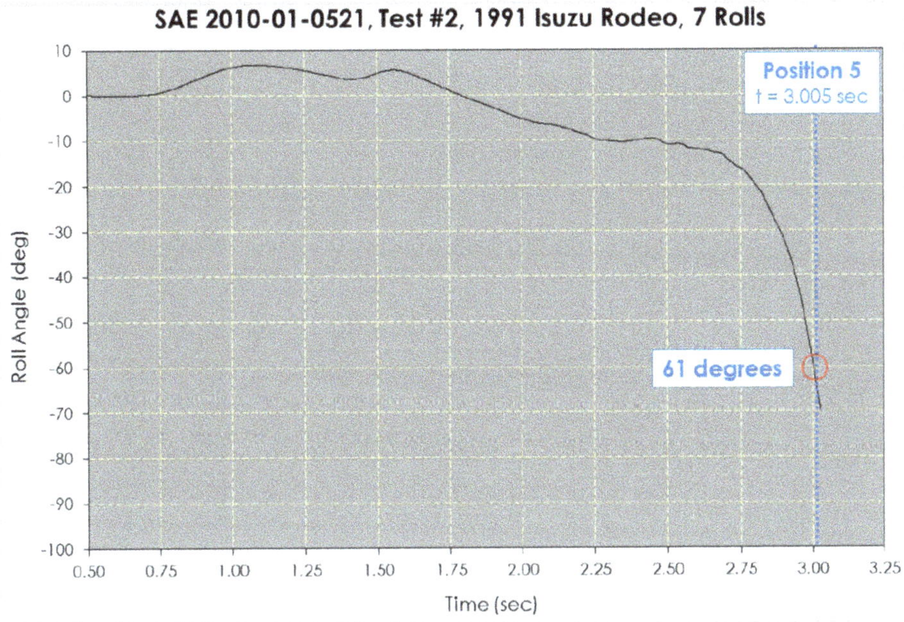

from 2.5 to 3.0 s. The roll velocity then levels off to an approximately constant value of 400°/s. This leveling off of the roll velocity is an indication that the trip phase has ended, the vehicle has become airborne, and the roll phase has begun. This plateauing of the roll velocity, which occurred at a time of approximately 3.005 s, was used to define the end of the trip phase. This point is identified in Figure 5.11.

Next, the roll velocity data was integrated to obtain the roll angle of the vehicle body throughout the first 3.25 s of the test. Figure 5.12 shows these roll angles. The roll angle builds up over an approximate time frame of 2.5 to 3.0 s, achieving a roll angle of approximately 61° at the end of the trip phase. Based on the determination that the trip phase ended 3.005 s after the vehicle was released, the video was reviewed and this time was identified. Figure 5.13 depicts a frame from the video showing the vehicle at the end of the trip phase. The video was also reviewed in conjunction with the physical evidence diagram of Figure 5.8 to identify Positions 1 (beginning of tire marks) and 4 (right front tire liftoff). These positions are shown in Figures 5.14 and 5.15. The video was not conducive to determining when the right rear tire lifted off.

Based on the timing of Positions 1, 4, and 5 in the video, these positions were identified on the graph of the lateral and longitudinal accelerations during the trip phase (Figure 5.16). The average lateral (ground plane) and longitudinal accelerations experienced by the vehicle were determined for the time frame from Positions 1 to 5 and for the time frame from Positions 4 to 5. It was concluded that from Positions 1 through 5 the vehicle experienced an average lateral acceleration of 1.09g and a longitudinal acceleration of 0.47g. Both of these were acting in a direction that would decelerate the vehicle. During the time frame from Positions 4 to 5, the vehicle experienced an average lateral acceleration of 1.69g and a longitudinal acceleration of 0.64g.

FIGURE 5.13 Frame from video showing Position 5 (end of trip).

FIGURE 5.14 Frame from video showing Position 1 (beginning of tire marks).

FIGURE 5.15 Frame from video showing Position 4 (right front tire liftoff).

FIGURE 5.16 Longitudinal and lateral accelerations during trip with key times labeled.

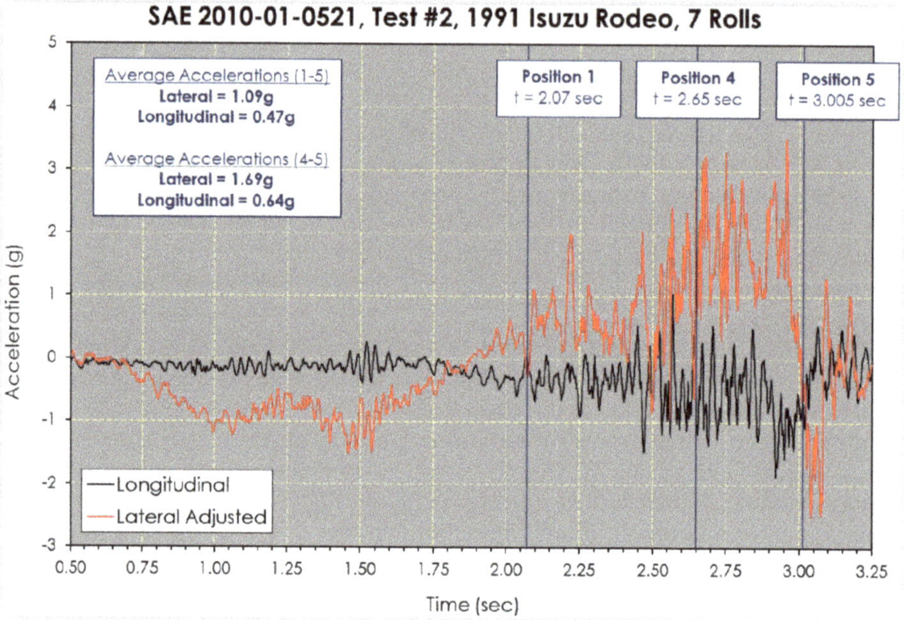

These measured accelerations demonstrate that there were longitudinal forces applied to the vehicle during the trip phase due to furrowing and steering effects (the significant rightward steer that was present throughout the trip phase is clearly visible in Figure 5.15). Such dynamic longitudinal forces have not customarily been included in analysis of the trip phase. Based on the analysis described in this section, however, their inclusion could potentially improve the accuracy of speed calculations related to the trip phase. Using the actual longitudinal accelerations, the speed analysis scenarios reported in Table 5.1 were recalculated and are reported in Table 5.2. The scenario that used Positions 3-5 as the trip phase used a longitudinal deceleration of 0.55g.

One issue with the calculations reported in Tables 5.1 and 5.2 is that they utilized a roll phase deceleration rate of 0.363g. This was the deceleration rate that Arndt reported for this test, and so, using this value allowed for the trip phase calculations to be isolated.

TABLE 5.2 Summary of trip phase calculations with actual longitudinal accelerations (assumes roll phase deceleration rate of 0.363g)

Trip phase	Actual speed at beginning of trip (mph)	Trip distance (ft)	Actual deceleration rate (g)	Calculated normalized lateral tripping force (g)	Average slip angle (°)	Calculated deceleration rate assuming actual longitudinal force	Calculated speed assuming actual longitudinal force (mph)
Positions 4-5	61.1	27.4	2.14	1.68	48	1.67	57.8 (−3.3)
Positions 3-5	64.3	45.5	1.58	1.38	44	1.35	61.7 (−2.6)
Positions 1-5	68.9	88.2	1.05	1.20	28	0.96	67.1 (−1.8)

TABLE 5.3 Summary of trip phase calculations with actual longitudinal accelerations (assumes roll phase deceleration rate of 0.44g)

Trip phase	Actual speed at beginning of trip (mph)	Trip distance (ft)	Actual deceleration rate (g)	Calculated normalized lateral tripping force (g)	Average slip angle (°)	Calculated deceleration rate assuming actual longitudinal force	Calculated speed assuming actual longitudinal force (mph)
Positions 4-5	61.1	27.4	2.14	1.68	48	1.73	61.8 (+0.7)
Positions 3-5	64.3	45.5	1.58	1.38	44	1.37	65.3 (+1.0)
Positions 1-5	68.9	88.2	1.05	1.20	28	0.96	70.3 (+1.4)

However, this roll phase deceleration rate is outside the range that would typically be employed by a reconstructionist. A more typical value would be 0.44g. Table 5.3 summarizes calculations similar to those reported in Table 5.2 (using the actual longitudinal decelerations), but this time using a roll phase deceleration rate of 0.44g. Using 0.44g results in overestimating the speed at the beginning of the roll phase, but with any definition of the trip phase, it improves the accuracy of the calculated speed over calculations that ignore the longitudinal forces.

Source of the Longitudinal Forces

A 2017 *Collision* article explored the hypothesis that the longitudinal forces applied to the Isuzu during the trip phase were due to the significant rightward steering input that was present during the trip phase [21]. PC-Crash crash simulation software was used for this analysis. This software allowed for easy incorporation of the effects of steering on the speed loss during the trip phase and also easy incorporation of the three-dimensional characteristics of the terrain. A three-dimensional terrain was used for the simulation, which was based on Asay's survey data. Manufacturer specifications provided the vehicle dimensions, and the vehicle weight reported by Asay was used (4,266 lb). Moments of inertia were estimated based on formulas in MacInnis [22] and Allen [23]. PC-Crash's TM-Easy tire model was used with the default parameters. Representative shape parameters for the vehicle body of the Isuzu were measured from a computer model and entered for the simulation.

To generate the simulation in PC-Crash, the vehicle was placed at Position 1 (Figure 5.8) and a speed of 68.9 mph was entered. The yaw rotational rate and velocity direction at Position 1 were varied during the simulation process to optimize the match of the subsequent vehicle motion with the tire marks and furrows. During the simulation process, the results were judged based on how closely the calculated vehicle motion followed the tire marks and furrows of the final clockwise yaw that preceded the roll phase.

FIGURE 5.17 Friction zones used in the PC-Crash simulation.

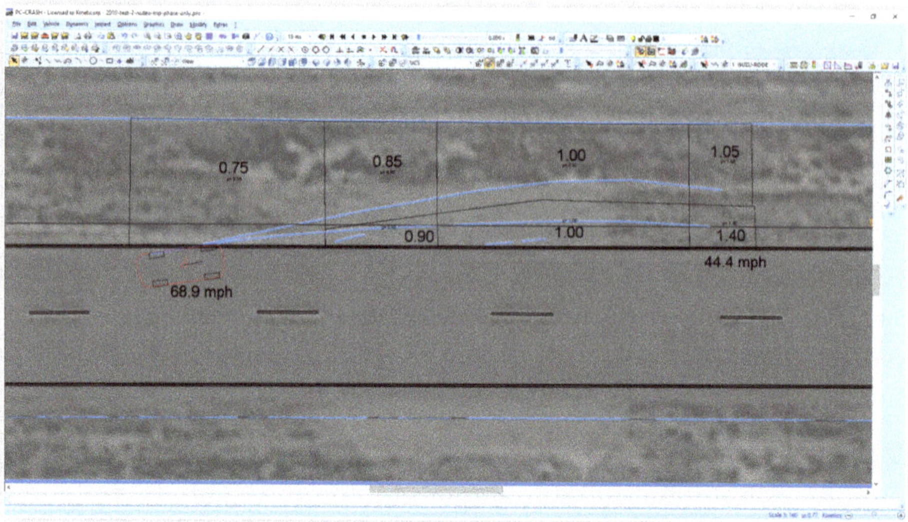

According to technical specifications for the Isuzu, the steering ratio was 18.7:1 and the minimum turning circle for the vehicle was 40.9 ft. For a rightward steering input, this implies a maximum input at the steering wheel of approximately 506°. This would produce a rightward steer at the left front tire of approximately 27° and a rightward steer at the right front tire of approximately 35°. This steering input was entered into PC-Crash for the simulation. Brake factors of 1% were entered for each tire of the Isuzu to represent rolling resistance. No additional braking was applied during the simulation. Friction zones were constructed for the trip phase in order to generate the known speed loss over the trip phase while also matching the tire marks and furrows and the known roll angle and roll angular velocity of the vehicle at the end of the trip. Figure 5.17 is a screen capture from the simulation that shows the friction zones that we utilized. This approach is similar to the one described by Grimes in a 2006 [24] study, though that study utilized HVE rather than PC-Crash.

Figure 5.18 shows the vehicle motion in the optimized simulation. In this simulation, the speed of the vehicle at the end of the trip phase was 44.56 mph, close to the

FIGURE 5.18 Vehicle motion in PC-Crash simulation.

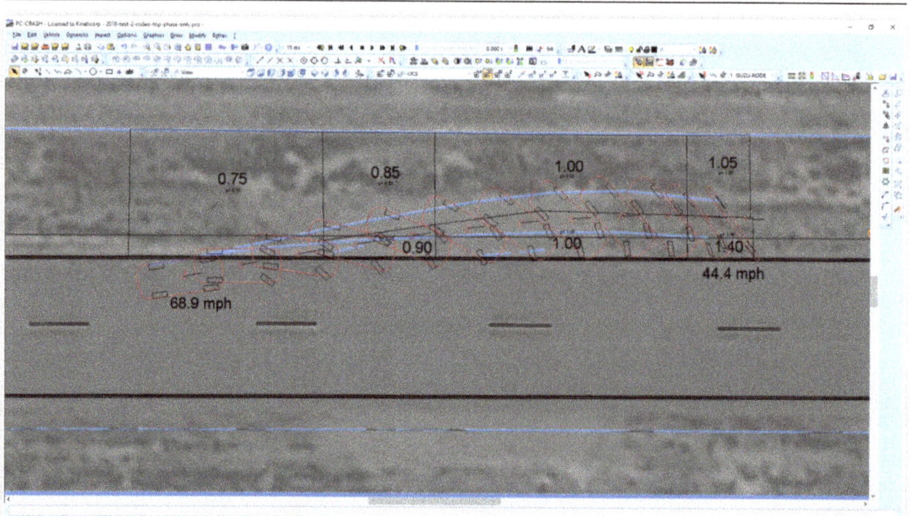

FIGURE 5.19 Vehicle motion in simulation without rightward steering input.

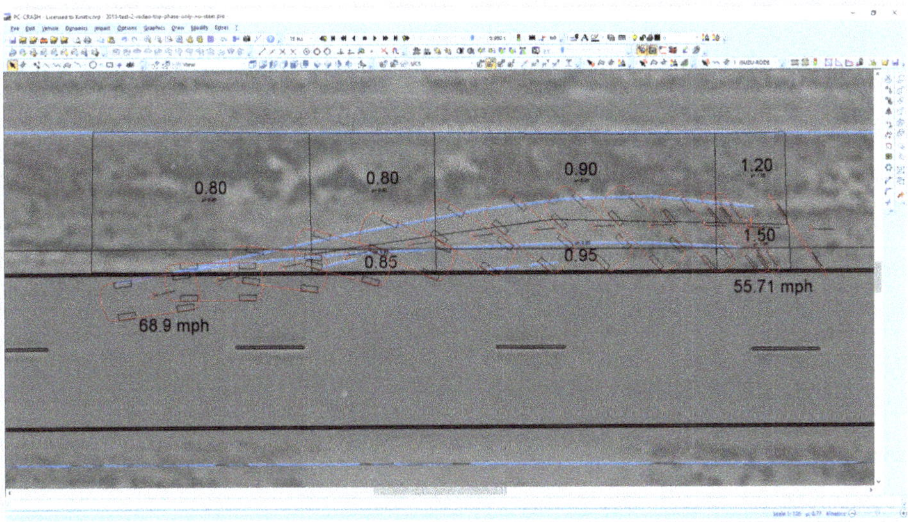

speed of 44.4 mph determined by Arndt. The roll angle at the end of trip was approximately 63.8°, and the roll velocity was 382°/s, reasonably consistent with the values determined from the sensor data. In the simulation, the average lateral acceleration during the final clockwise yaw rotation of the vehicle was 1.24g and the longitudinal acceleration was 0.36g, both of which acted to decelerate the vehicle.

After optimizing this simulation with the rightward steering input, a simulation was generated with the rightward steering input removed. Without this steering input, we were not able to generate a simulation with a satisfactory and simultaneous match with the tire marks, the furrows, the vehicle speeds, and the roll angle and roll angular velocity at the end of trip. One attempt is shown in Figure 5.19. In this simulation, the match with the tire marks and furrows was acceptable, though not perfect. The speed loss during the trip, however, was significantly underestimated with the speed entering the roll phase being more than 11 mph too high. This indicates a significant drop in the overall deceleration level from the simulation that included the rightward steering input. The roll velocity at the end of the trip phase was also significantly underestimated at 230°/s, rather than the actual 400°/s.

The longitudinal accelerations in the simulation without steering were significantly less than in the previous simulation. The magnitude of the longitudinal accelerations in this simulation were consistent with the level that would be expected from rolling resistance and from traversing the slope of the roadway shoulder during the trip (less than 0.1g), but they were not anywhere near the level that actually existed during the test. Together, the simulation with steering and the one without steering confirmed that the longitudinal accelerations observed in the sensor data of the actual test were largely, though not completely, due to the significant rightward steering input that was present during the trip phase.

This finding indicates that there is not a need to develop any new models or formulas in order to incorporate these longitudinal forces into modeling of the trip phase. Accident reconstructionists already have a number of methods for incorporating the effects of steering on the deceleration rate of a yawing vehicle. Simulation is one of these [25]. Carter [26] discussed several other calculation techniques that would allow a reconstructionist to calculate the effects of steering on the deceleration rate of a yawing vehicle. One difficulty would be knowing which direction, and the extent to which, a driver was steering during the trip phase. This is where simulation has a distinct advantage over spreadsheet calculations. In modeling the trip phase of the Rodeo test, the match with the physical evidence was significantly improved by including the rightward steering input. This is an indication that simulation can help detect the presence of a steering input during the trip phase. In cases where a vehicle begins yawing on asphalt before it enters the soil, striations in the tire marks on the asphalt may also be useful in determining which direction and the extent to which the driver was steering during the trip phase [27, 28]. In some instances, data from an event data recorder might also give the reconstructionist knowledge of the steering inputs.

On-Road Rollovers

So far, this chapter has covered trip modeling appropriate for soil-tripped and curb-tripped rollovers. For calculating the speed loss during the trip phase, Equations (5.8) and (5.9) have been the centerpiece of the discussion so far. At this point, it is important to note that Equation (5.8) should, in most cases, not be applied to determine the speed loss for a rollover that is initiated on the roadway. Equation (5.8) always yields a normalized lateral force that is greater than a vehicle's static stability factor. However, the lateral forces generated between a vehicle's tires and an asphalt or concrete roadway surface is usually lower than the static stability factor. This means that, the lateral forces are not alone sufficient to initiate a rollover on the road surface. Suspension effects due to severe steering inputs are also playing a role in initiating the rollover. When a rollover occurs due to a combination of the lateral tire forces and suspension effects due to severe steering inputs, Equation (5.8) can overestimate the lateral forces, and thus, the speed loss during the roll initiation phase. In these instances, the roll initiation phase can be lumped into the loss-of-control phase, and the speed loss calculated is the same way it is for the rest of the loss-of-control phase [29].

References

1. Martinez, J. and Schlueter, R., "A Primer on the Reconstruction and Presentation of Rollover Accidents," SAE Technical Paper 960647, 1996, doi:10.4271/960647.
2. Orlowski, K., Moffatt, E., Bundorf, R., and Holcomb, M., "Reconstruction of Rollover Collisions," SAE Technical Paper 890857, 1989, doi:10.4271/890857.
3. Asay, A. and Woolley, R., "Rollover Testing of Sport Utility Vehicles (SUVs) on an Actual Highway," SAE Technical Paper 2010-01-0521, 2010, doi:10.4271/2010-01-0521.
4. Larson, R., Smith, J., Werner, S., and Fowler, G., "Vehicle Rollover Testing, Methodologies in Recreating Rollover Collisions," SAE Technical Paper 2000-01-1641, 2000, doi:10.4271/2000-01-1641.
5. Garrott, W., Howe, J., and Forkenbrock, G., "Results from NHTSA's Experimental Examination of Selected Maneuvers That May Induce On-Road Untripped, Light Vehicle Rollover," SAE Technical Paper 2001-01-0131, 2001, doi:10.4271/2001-01-0131.
6. Wilson, L.A., Gilbert, M., and Godrick, D., "Reconstruction and Analysis of Steering-Induced, On-Road, Untripped SUV Rollover Tests (Part 1)," *Collision-The International Compendium for Crash Research* 2, no. 1, Spring 2007.

7. Wilson, L.A., Gilbert, M., and Godrick, D., "Reconstruction and Analysis of Steering-Induced, On-Road, Untripped SUV Rollover Tests (Part 2)," *Collision-The International Compendium for Crash Research* 2, no. 2, Fall 2007.
8. Stevens, D., Arndt, S., Wayne, L., Arndt, M. et al., "Rollover Crash Test Results: Steer-Induced Rollovers," SAE Technical Paper 2011-01-1114, 2011, doi:10.4271/2011-01-1114.
9. Marine, M., Wirth, J., and Thomas, T., "Characteristics of On-Road Rollovers," SAE Technical Paper 1999-01-0122, 1999, doi:10.4271/1999-01-0122.
10. Cooperrider, N., Thomas, T., and Hammoud, S., "Testing and Analysis of Vehicle Rollover Behavior," SAE Technical Paper 900366, 1990, doi:10.4271/900366.
11. Cooperrider, N., Hammoud, S., and Colwell, J., "Characteristics of Soil-Tripped Rollovers," SAE Technical Paper 980022, 1998, doi:10.4271/980022.
12. Asay, A. and Woolley, R., "Rollover Testing on an Actual Highway," *SAE Int. J. Passeng. Cars - Mech. Syst.* 2, no. 1 (2009): 1237-1267, doi:10.4271/2009-01-1544.
13. Jones, I., "The Mechanics of Rollover as the Result of Curb Impact," SAE Technical Paper 750461, 1975, doi:10.4271/750461.
14. Hughes, R., Lewis, L., Hare, B., Ishikawa, Y. et al., "A Dynamic Test Procedure for Evaluation of Tripped Rollover Crashes," SAE Technical Paper 2002-01-0693, 2002, doi:10.4271/2002-01-0693.
15. Brach, R.M. and Brach, R.M., *Vehicle Accident Analysis and Reconstruction Methods*, (Warrendale, PA: SAE, 2005), ISBN:978-0-7680-3437-0.
16. Rose, N., Beauchamp, G., and Fenton, S., "Factors Influencing Roof-to-Ground Impact Severity: Video Analysis and Analytical Modeling," SAE Technical Paper 2007-01-0726, 2007, doi:10.4271/2007-01-0726.
17. Gopal, M., Baron, K., and Shah, M., "Simulation and Testing of a Suite of Field Relevant Rollovers," SAE Technical Paper 2004-01-0335, 2004, doi:10.4271/2004-01-0335.
18. Shi, Y. and Nusholtz, G., "Simple Models for Analysis of Curb- and Soil-Trip Rollover Events," SAE Technical Paper 2006-01-0722, 2006, doi:10.4271/2006-01-0722.
19. Jones, I.S., "Road Accident Studies with Particular Reference to the Handling and Stability Characteristics of Cars," Ph.D. thesis, University of London, London, U.K., 1973.
20. Lund, Y.I. and Bernard, J.E., "Analysis of Simple Rollover Metrics," SAE Technical Paper 950306, 1995, doi:10.4271/950306.
21. Rose, N. and Carter, N., "An Analytical Review of Two Decades of Research Related to PC-Crash Simulation Software," SAE Technical Paper 2018-01-0523, 2018, doi:10.4271/2018-01-0523.
22. MacInnis, D., Cliff, W., and Ising, K., "A Comparison of Moment of Inertia Estimation Techniques for Vehicle Dynamics Simulation," SAE Technical Paper 970951, 1997, doi:10.4271/970951.
23. Allen, R., Klyde, D., Rosenthal, T., and Smith, D., "Estimation of Passenger Vehicle Inertial Properties and Their Effect on Stability and Handling," SAE Technical Paper 2003-01-0966, 2003, doi:10.4271/2003-01-0966.
24. Grimes, W., Balasa, J., Hunter, E., and Stevens, D., "Analyzing The Trip-Phase of Soft-Soil Rollovers," SAE Technical Paper 2006-01-1558, 2006, doi:10.4271/2006-01-1558.

25. Rose, N., Carter, N., and Beauchamp, G., "Post-Impact Dynamics for Vehicles with a High Yaw Velocity," SAE Technical Paper 2016-01-1470, 2016, doi:10.4271/2016-01-1470.

26. Carter, N., Beauchamp, G., and Rose, N.A., "Comparison of Calculated Speeds for a Yawing and Braking Vehicle to Full-Scale Vehicle Tests," SAE Technical Paper, 2012-01-0620, 2012, doi:10.4271/2012-01-0620.

27. Beauchamp, G., Hessel, D., Rose, N., Fenton, S. et al., "Determining Vehicle Steering and Braking from Yaw Mark Striation," *SAE Int. J. Passeng. Cars - Mech. Syst.* 2, no. 1, (2009): 291-307, doi:10.4271/2009-01-0092.

28. Beauchamp, G., Pentecost, D., Koch, D., and Rose, N.A., "The Relationship between Tire Mark Striations and Tire Forces," *SAE Int. J. Trans. Safety* 4, no. 1 (2016): 134-150, doi:10.4271/2016-01-1479.

29. Beauchamp, G., Thornton, D., Bortles, W., and Rose, N.A., "Tire Mark Striations: Sensitivity and Uncertainty Analysis," *SAE Int. J. Trans. Safety* 4, no. 1 (2016): 121-127, doi:10.4271/2016-01-1468.

Appendix

FIGURE 5.20 Vertical velocity versus time, linearly increasing trip force (left) and sinusoidal trip force (right) [SI units].

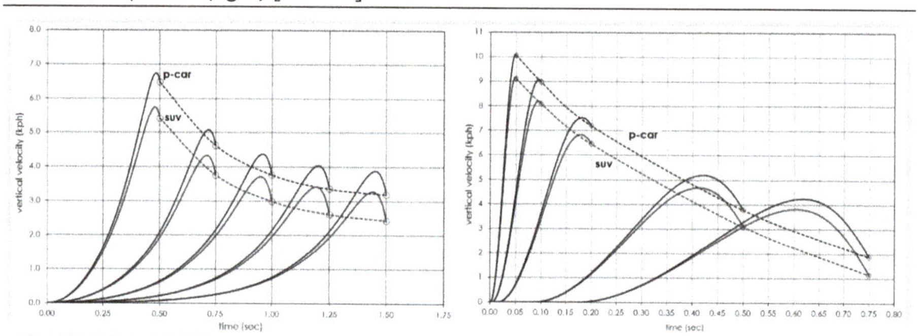

CHAPTER 6

Analysis Methods: Loss-of-Control Phase

To calculate the speed of a vehicle at the beginning of the loss-of-control phase, first reconstruct the motion of the vehicle during this phase (typically based on tire marks), and then determine the deceleration rate associated with that motion. This chapter focuses on models that yield the vehicle's deceleration rate during this phase, including critical speed analysis.

Determining the rate at which the vehicle decelerates during the loss of control depends on having a model that can deal with the combined effects of yawing and braking. In this chapter, three approaches to determining this deceleration rate is discussed [1]. The first approach uses the Bakker-Nyborg-Pacejka (BNP) tire force model in conjunction with the Nicolas-Comstock-Brach (NCB) combined tire force equations to calculate a yawing and braking vehicle's deceleration rate. The second approach is a simpler equation for calculating a yawing and braking vehicle's deceleration rate that was proposed by Martinez and Schlueter [2]. It will be shown that this equation exhibits physically unrealistic behavior, and thus, it should not be used for reconstruction calculations. The third approach uses an equation that is attributed to the CRASH program in the 2010 edition of *Traffic Crash Reconstruction* by Lynn Fricke. This equation is nearly as simple as the Martinez and Schlueter equation but its behavior is more reasonable. Comparisons are made to two full-scale vehicle tests, both involving yawing and braking vehicles.

Bakker-Nyborg-Pacejka (BNP) and Nicolas-Comstock-Brach (NCB) Model

This first method begins by modeling and calculating normalized longitudinal and lateral tire forces as if they could be decoupled, calculating each one as if the other was not present. The braking (or longitudinal slip) level is used to calculate a normalized longitudinal tire force component as if there was no sideslip angle. Then, the tire sideslip angle is used to calculate a normalized lateral tire force component as if there was no braking. In this chapter, these calculations are carried out using a zero-sideslip angle longitudinal tire force curve and a free-rolling lateral tire force curve generated using generic, representative coefficient values with the "magic formula" of Bakker, Nyborg, and Pacejka - the BNP model [3]. The BNP model equations are used to calculate decoupled normalized longitudinal and lateral tire forces as a function of the longitudinal wheel slip ratio (s) and normalized tire sideslip angle ($2\alpha/\pi$). The symbol α refers to the tire sideslip angle (measured in radians).

To determine the vehicle deceleration rate, there is not any need to calculate the actual magnitude of the tire forces. Instead, the calculations discussed here yield tire forces that have been normalized with the normal load on the tire. The letter Q is used to refer to these normalized tire forces. Equation (6.1) gives the generic functional form of the BNP equations. Both the lateral and longitudinal tire force equations take this form. The variable u is a generic slip ratio that can be either the longitudinal slip ratio, s, or the lateral sideslip ratio, $2\alpha/\pi$.

$$Q(u) = D\sin\left\{C\tan^{-1}\left[B(1-E)Ku + E\tan^{-1}(BKu)\right]\right\} \tag{6.1}$$

When using this function to model the normalized longitudinal tire force, $Q_x(s)$, the longitudinal slip ratio of the tire, s, can vary between 0 and 1. A value of 0 corresponds to a free-rolling tire and a value of 1 corresponds to a fully locked tire. Similarly, for the normalized lateral tire force, $Q_y(2\alpha/\pi)$, the lateral sideslip ratio or normalized tire sideslip angle can vary between 0 and 1. A value of 0 corresponds to a sideslip angle of zero, and a value of 1 corresponds to a sideslip angle of 90°. The parameter K in Equation (6.1) acts as a scaling value that can be set to a value of 100 to convert the lateral and longitudinal slip ratios to percentages. The coefficients B, C, D, and E are shape parameters that are chosen to empirically model a specific set of tire data. For the calculations described here, the shape parameters were chosen to yield typical or representative tire force curves based on data in Brach [4]. Figure 6.1 depicts the longitudinal (left) and lateral (right) tire force curves that were generated. Table 6.1 lists the coefficients used to generate these graphs.

FIGURE 6.1 Typical longitudinal and lateral tire force curves.

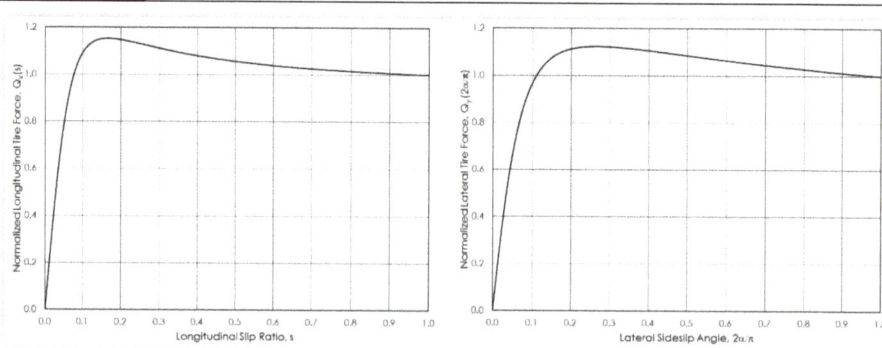

For a vehicle undergoing both yawing and braking, the magnitudes of the longitudinal and lateral components of the tire force cannot be decoupled in the manner depicted in Figure 6.1. For instance, braking reduces the force available for cornering. To calculate an actual resultant tire force, these decoupled tire forces need to be combined and resolved in a way that incorporates their interdependence and effect on one another. The NCB model provides such a method for combining decoupled longitudinal and lateral tire forces [4, 5, 6, 7]. These equations are listed below.

TABLE 6.1 Coefficient values for lateral and longitudinal BNP equations

	Value for longitudinal tire force curve	Value for lateral tire force curve
B	0.118	0.093
C	1.398	1.498
D	1.152	1.124
E	−0.128	0.573

$$Q_x(\alpha,s) = \frac{Q_x(s)Q_y(\alpha)}{\sqrt{s^2Q_y^2(\alpha)+Q_x^2(s)\tan^2\alpha}} \cdot \frac{\sqrt{s^2C_\alpha^2+(1-s)^2\cos^2\alpha Q_x^2(s)}}{C_\alpha} \tag{6.2}$$

$$Q_y(\alpha,s) = \frac{Q_x(s)Q_y(\alpha)}{\sqrt{s^2Q_y^2(\alpha)+Q_x^2(s)\tan^2\alpha}} \cdot \frac{\sqrt{(1-s)^2\cos^2\alpha Q_y^2(\alpha)+\sin^2\alpha C_s^2}}{C_s\cos\alpha} \tag{6.3}$$

C_S and C_α are determined from the coefficients in the BNP equations and are defined as follows:

$$C_s = (BCDK)_s \tag{6.4}$$

$$C_\alpha = (BCDK)_\alpha \tag{6.5}$$

Once the normalized components of the tire forces are calculated, a deceleration rate for each tire can be calculated with the following equation [5].

$$f_i = \mu_0 \cdot \left[Q_x(\alpha,s)\cdot\cos(\alpha)+Q_y(\alpha,s)\cdot\sin(\alpha)\right] \tag{6.6}$$

This equation projects the tire force components onto the vehicle velocity direction to determine the portion of these tire forces that is opposing the velocity at any particular point. In this equation, μ_0 is the roadway coefficient of friction and f_i is the effective deceleration rate at the *i*th wheel position. In calculating the deceleration rate for the vehicle, an effective deceleration rate would be calculated for each wheel position. Once an effective deceleration rate is obtained for each wheel position, they can be combined into an effective deceleration rate for the vehicle as a whole with the following equation. In this equation, W_{total} is the total weight of the vehicle, W_i is the static weight on each wheel position, and N is the total number of wheels.

$$f_{veh} = \frac{1}{W_{total}}\sum_{i=1}^{N}\left[f_i \cdot W_i\right] \tag{6.7}$$

Brach's papers have demonstrated that when combined, the BNP and NCB models can yield realistic tire force component calculations and vehicle motion for simulations with conditions of combined braking and cornering [4, 5, 6, 7]. Later, this chapter examines results that verify that this accuracy translates from a forward-simulation context to a backward-reconstruction context.

Martinez-Schlueter Model

Martinez and Schlueter [2] proposed the following equation for calculating the deceleration rate of a vehicle undergoing both yawing and braking:

$$f_{veh} = \mu_0 \sin\left[\alpha + \tan^{-1}\left(\frac{\kappa_B}{\sqrt{1-\kappa_B^2}}\right)\right] \tag{6.8}$$

In this equation, κ_B is a parameter Martinez and Schlueter termed the fractional lockup. This term is not defined in their paper, but presumably it was intended to equate to the braking level as a proportion of the available friction, with values ranging from 0 to 1. This model does not behave in a physically realistic way and it should not be used. It is presented here in order to be thorough and to explain why other modeling approached would be preferred.

As an example, consider a vehicle with a sideslip angle of 90°. At this sideslip, the deceleration rate of the vehicle should be equal to the friction coefficient, independent of the level of braking. However, for a 90° sideslip angle and a fractional lockup of 0.5, Equation (6.11) yields a deceleration rate that is approximately 87% of the friction coefficient. The deceleration rate declines as the braking level increases, approaching zero as the fractional lockup approaches 1.0 (locked-wheel braking).

Examination of the bracketed terms in Equation (6.8) reveals the source of this error. As κ_B approaches a value of 1.0, the inverse tangent term approaches 90 degrees. When this angle is combined with the sideslip angle, the result is a term that can vary between 0 and 180°. Because the sine function decreases for angles greater than 90°, Equation (6.8) begins decreasing anytime the bracketed angle term exceeds 90°. This means that for a fractional lockup of 1.0, Equation (6.8) exhibits particularly bad behavior since it starts at a deceleration equal to the friction coefficient at a slip angle of 0 (correct), but then decrease to 0 as the slip angle reaches 90 degrees (the opposite of correct).

CRASH Model

Equation (6.9) gives another option for calculating a yawing and braking vehicle's deceleration rate. In the 2010 edition of *Traffic Crash Reconstruction* by Lynn Fricke, this equation is attributed to the CRASH program.

$$f_{veh} = \mu_0 \sqrt{\sin^2 \alpha + \left(\frac{f_{long}}{\mu_0}\right)^2 \cos^2(\alpha)} \tag{6.9}$$

In this equation, f_{long} is the longitudinal deceleration rate that would be achieved were there no side slip. The value of f_{long} can vary between 0 and μ_0. With the application of this equation, it should be noted that f_{long} is different than the longitudinal slip. In practice, the ratio of f_{long} to μ_0 can be determined from slip using a normalized longitudinal tire force plot, such as that in Figure 6.1. For example, if the tire slip was 5%, this ratio would equal approximately 0.8. Typically, the full available friction for longitudinal deceleration would be recruited by the time the longitudinal slip reaches 20 to 30%.

When the braking level is zero, Equation (6.9) reduces to the following:

$$f_{veh} = \mu_0 \cdot \sin\alpha \tag{6.10}$$

FIGURE 6.2 Normalized deceleration rate from the CRASH equation.

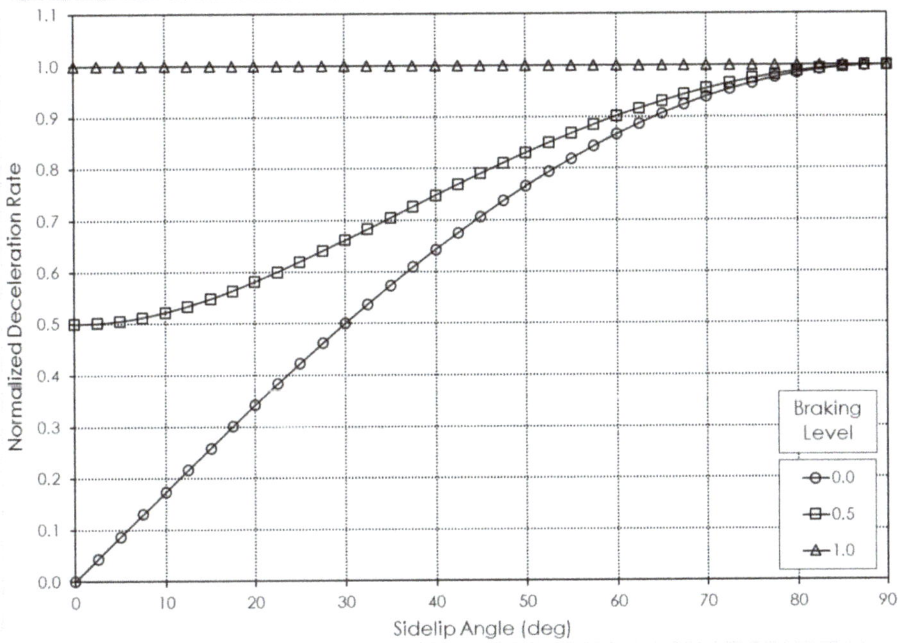

When the longitudinal deceleration rate reaches a maximum (full braking), Equation (6.9) reduces to the following:

$$f_{veh} = \mu_0 \tag{6.11}$$

The behavior and physical accuracy of Equation (6.9) is significantly better than that of Equation (6.8). Sample curves of the square root term (the normalized deceleration rate) from this equation are plotted in Figure 6.2. In this graph, the normalized deceleration rate is plotted as a function of sideslip angle and braking level. As this graph shows, Equation (6.8) yields a normalized deceleration rate of 1.0 for a braking level of 1.0, independent of the sideslip angle – physically correct behavior. For braking levels less than 1.0, the equation yields deceleration rates that start out at the braking level when the sideslip angle is zero and build up to 1.0 as the sideslip angle goes to 90 degrees, which is again physically reasonable behavior.

Comparison to Full-Scale Tests

This section compares results from the BNP/NCB and CRASH models to the results of two full-scale vehicle tests, both involving a yawing and braking 2008 Chevrolet Malibu outfitted with Hankook Winter Ipike tires (225/60R16). The photograph in Figure 6.3 depicts this vehicle in the midst of one of these tests. This Malibu was instrumented with a VBOX data acquisition system manufactured by Racelogic. The VBOX recorded the vehicle's translational and angular position, and the Vehicle CAN Interface recorded wheel speeds and steering position from the Malibu's internal sensors. All data were recorded at 20 Hz. The GPS antenna was magnetically attached to the roof of the Malibu. In addition to the two yaw tests, a straight-line skid-to-stop test was run to quantify the coefficient of friction for the asphalt test surface. For this test, the ABS was disabled to allow the wheels to lock. This test yielded a sliding friction value of 0.79.

FIGURE 6.3 Vehicle in the midst of yaw Test #2.

Test #1: In the first test, the Malibu was accelerated to a speed of approximately 48 mph (77 kph). The driver applied the parking brake and made a steering input to the left to induce yaw rotation. Once the yaw was initiated, the driver released the parking brake and the vehicle yawed approximately 180° in a counterclockwise direction. Once the parking brake was released, the driver reported no additional braking. The wheel speed sensor data, though, indicated there may have been a small amount of braking near the end of the test, when the vehicle was rolling backwards to rest.

The position of the test vehicle was documented before and after the test. The test surface was surveyed and photographed, as were tire marks that the vehicle deposited during the test. A diagram was created from this survey data. The data acquired with the VBOX was synchronized to the tire marks by importing both the evidence diagram and the VBOX data into an animation software package (Autodesk MAX). The VBOX data was assigned to a vehicle model within that animation software package, and the motion of this vehicle was aligned to the tire marks. Once synchronized in this manner, the VBOX data could be examined at each vehicle location of interest.

The tire marks were analyzed to obtain the longitudinal slip and sideslip angle for each of the Malibu's tires at a number of positions along the vehicle path. The sideslip angles for the rear tires were determined based on the alignment of the vehicle with the tire marks. Since the front tires are steerable, the sideslip angles of the front tires were determined by adding or subtracting the steering angle at the tire from the sideslip angles at the rear tires. The longitudinal slip percentages were then calculated using striation analysis. Techniques for analyzing striations in tire marks are discussed in the next chapter. Suffice it to say, for now, that equations can be written that relate the orientation and spacing of yaw mark striations to the vehicle braking and steering levels present at the time the striations were deposited. Conceptually, such analysis relies on the idea that tire mark striations are direct physical evidence of the actual direction of the force that was applied to the tire at the time the mark was deposited. Without braking, the tire force is perpendicular to the tire heading, and thus, the striations produced are perpendicular to the tire heading.

Test #2: In the second test, the Malibu was accelerated up to a speed of approximately 48 mph (77 kph). The parking brake was then applied and the driver made a steering input to the right initiating a clockwise yaw. The parking brake was then released, and the driver applied the service brakes and countersteered aggressively to the left. During the event, the vehicle yawed approximately 90°. The wheel speed sensor data obtained from the VBOX showed that the ABS did not engage during the test.

Again, the position of the test vehicle was surveyed both before and after the test. The test surface and the tire marks that resulted from the test were surveyed and photographed. A diagram was created from this survey data and the data acquired with the VBOX was synchronized to the tire marks. The tire marks were analyzed to obtain inputs for the deceleration rate calculations [8]. Some of the results of this tire mark analysis are shown in the graph below (Figure 6.4). This graph shows the longitudinal slip percentages for the left rear wheel. At the first position, the driver was in the process of releasing the parking brake (60% slip). The next three positions have longitudinal slip near 0% and represent the time between the release of the parking brake and application of the service brakes. Reflecting the service brake application, the slip percentage then increases to a value of around 10% for the 5th and 6th positions.

For each of these tests, speed analysis was conducted using the BNP/NCB model and CRASH model. Using diagrams created from surveys of the tire marks from each test, vehicle models were aligned with the tire mark evidence as depicted in Figures 6.5 and 6.6 (for clarity, these graphics do not depict all of the positions used in the analysis). The segment lengths between positions were measured and the average vehicle sideslip angles were calculated based on the angles at the beginning and end of the segments. To account for braking, the longitudinal slip percentages determined from striation analysis were used in the BNP/NCB and CRASH models.

For the portions of the two tests during which the parking brake was applied, the rear tires of the Malibu locked while the front tires were unaffected. For portions of the tire marks where this lockup was apparent, the longitudinal slip percentage of the rear tires was assigned a value of 100%. Once the parking brake was released, striations from the rear-wheel tire mark from the leading side of the vehicle were used to calculate

FIGURE 6.4 Results of tire mark analysis for Test #2.

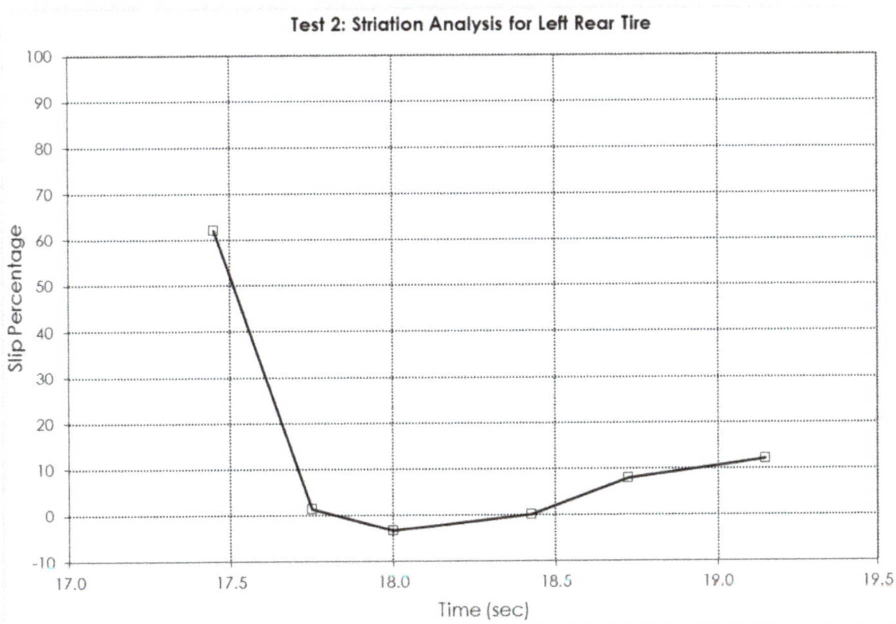

FIGURE 6.5 Vehicle model matched to tire mark evidence for Test #1.

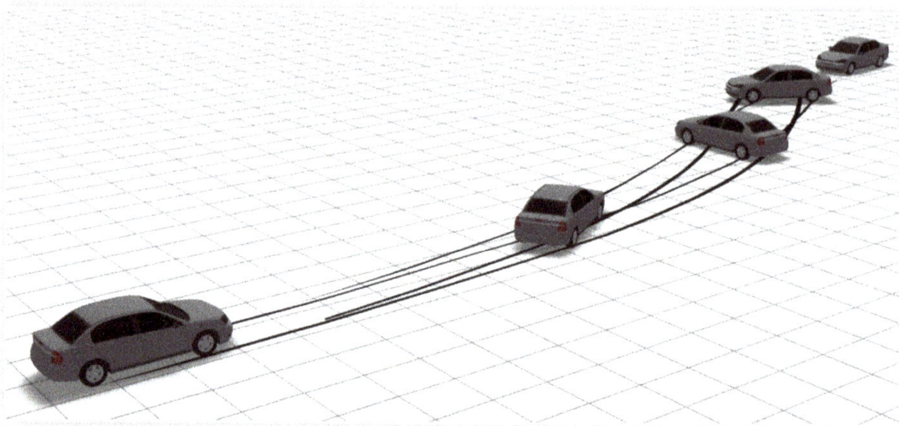

FIGURE 6.6 Vehicle model matched to tire mark evidence for Test #2.

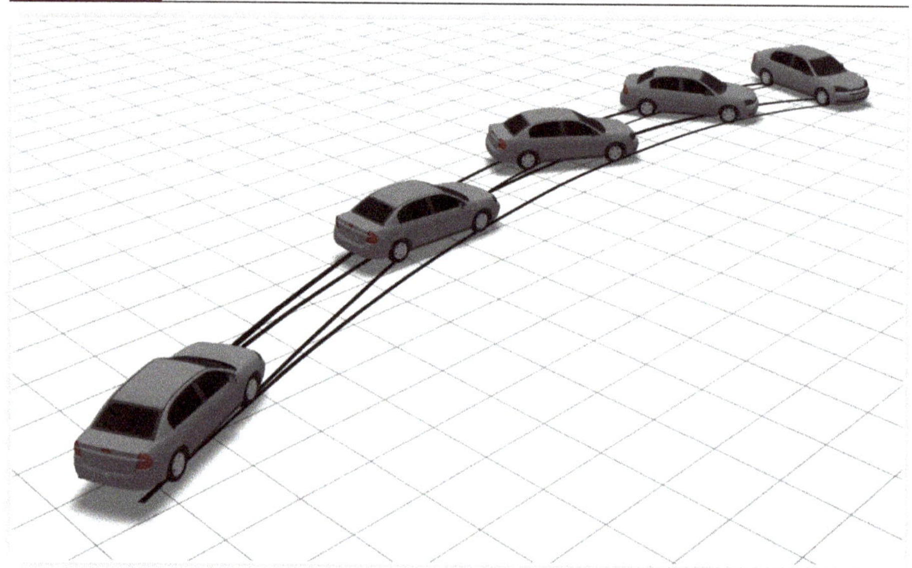

longitudinal slip percentages, and these values were applied to all tires. When the striation analysis indicated that a driven wheel was free rolling, the wheel was assigned a longitudinal slip percentage of 0.5%, to correspond with the measured rolling resistance of a similar vehicle documented by Cliff [9]. When tire marks from a non-driven wheel indicated no braking, a longitudinal slip value of 0% was used. In Test 1, the service brakes were not applied after the parking brake was released, so the longitudinal slip percentages were assigned values commensurate with free-rolling wheels. In Test #2, we used the longitudinal slip percentages shown in Figure 6.4.

Once the deceleration rate for each segment was determined, the vehicle speeds over the course of the yaws could be calculated with basic kinematic equations. Figures 6.7 and 6.8 depict the results of this analysis (versions of these graphs in SI units are included in the Appendix of this chapter). In these graphs, vehicle speed is plotted against distance. The dotted black line represents the vehicle speed obtained from the VBOX. The unfilled square markers represent the speeds obtained with the BNP/NCB model, and the filled circle markers represent the speeds obtained with the CRASH model. Both the BNP/NCB model and CRASH model yielded acceptable estimates for the vehicle speeds.

FIGURE 6.7 Speed analysis results for Test #1.

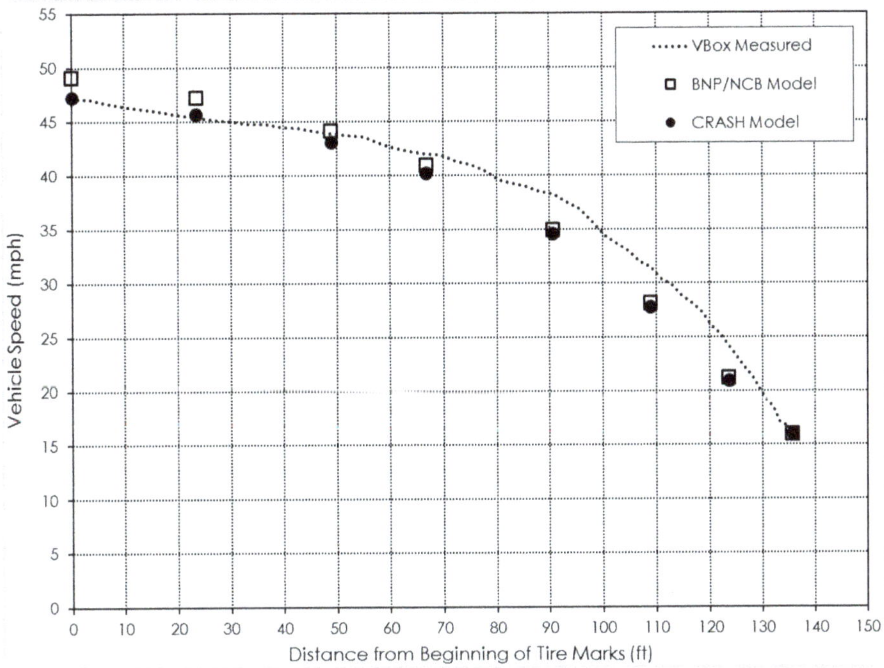

FIGURE 6.8 Speed analysis results for Test #2.

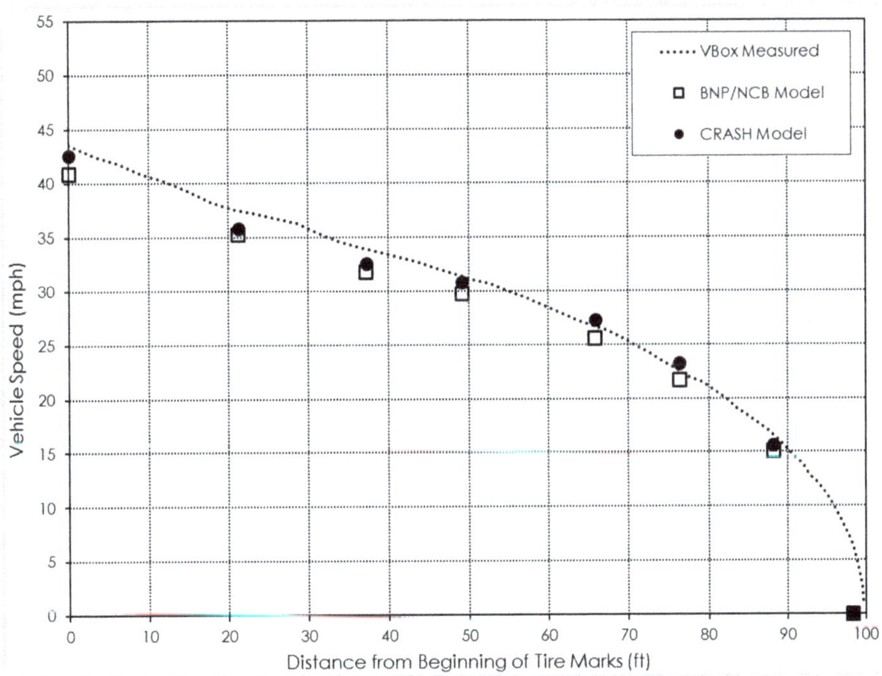

Critical Speed Analysis

Critical speed analysis offers another option for calculating the vehicle speed during the loss-of-control phase. The critical speed is the speed at which a vehicle following a curved path of radius R utilizes all of the available friction and begins sliding to the outside of the curve. The available friction is characterized with the friction coefficient μ. When all of the available friction has been recruited, tire marks are typically deposited and the radius of the vehicle path can be determined based on these tire marks. The critical speed formula is given by Equation (6.12). Sledge [10, 11] presents the derivation of this formula.

$$v_{critical} = \sqrt{\mu \cdot g \cdot R} \tag{6.12}$$

The critical speed formula is derived with the following assumptions:

1. The vehicle travels a *constant speed* while it traverses a *constant radius path*. In theory, this translates to assuming the vehicle is not decelerating due to sideslip or braking. In practice, though, the formula has been tested and validated for determining the speed of vehicles experiencing mild levels of sideslip and braking [12].

   Constant Speed: A vehicle can lose speed during a yaw due to the driver braking or due to the development of sideslip. As regards braking, Daily [IPTM] [13] notes that before applying critical speed analysis, the reconstructionist should ensure that "on paved surfaces, the marks contain diagonal striations indicating the tire was simultaneously rolling and slipping laterally. Without the diagonal striations, the mark could simply be a curved skid mark." In other words, for vehicles without antilock brakes, the presence of lateral striations in the tire mark ensures that the driver was not applying a significant level of braking while the tire marks were being deposited. Antilock braking systems can keep even heavily braked wheels at a low enough slip level that lateral striations can be deposited. Other authors confirm the importance of the presence of striations in the tire marks in order for critical speed analysis to yield accurate results [14, 15].

   Constant Radius Path: Masory [16] examined the validity of the circular path assumption and concluded that this assumption was reasonable in the testing he conducted. Clearly, though, the tire marks for any particular case can demonstrate the degree to which this assumption is valid for that specific case. Daily noted (contrary to the theoretical assumption of the model) that "a vehicle usually undergoes longitudinal deceleration during a [critical speed yaw] maneuver. As a result, the radius of the tire marks will decrease through the maneuver. This change in radius is detected by measuring the curved marks in two different places."

2. The vehicle is modeled as a point mass. In practice, this assumption translates to assuming the vehicle has not developed a significant sideslip angle (which is also related to the constant speed and constant radius assumptions). Summarizing other authors, Richardson [15] indicates that the sideslip angle should not exceed 25° over the distance the radius measurement is taken. Daily states that the sideslip angle should not exceed 20° over the distance that two radius measurements are taken. Daily's criteria imply that the slip angle should not exceed 20° for at least the first 60 ft of the yaw in order for the critical speed formula to be applied.

3. Tire marks are deposited because the vehicle has begun to slide to the outside of the curve, having reached the maximum lateral acceleration level that can be resisted by the available friction. One potential pitfall of critical speed analysis is that an analyst might apply the critical speed equation to tire marks deposited at a lateral acceleration level below the available friction. Testing by Baumann [17] demonstrated that tire marks can be deposited at acceleration levels below the peak friction, though the tire mark intensity appeared to be related to the acceleration level. Testing reported by Brown [18] and Seipel [19] also demonstrated that tire mark intensity is slip-level dependent, such that higher acceleration levels should generally produce darker tire marks. Baumann also noted that "tire design, construction and tire pressure are fundamental parameters influencing tire mark properties."

Application of the critical speed formula requires the reconstructionist to measure the radius of the path followed by the vehicle and to estimate the coefficient of friction. Fricke [14] and Daily [IPTM] [13] describe techniques for measuring the radius of the vehicle path. Fricke accepts the possibility of determining the radius from a scaled diagram, but Daily states, "Tire mark evidence *must* be gathered at the scene to be used in [critical speed analysis]. Estimating the radius based on indirect techniques, such as measuring CAD drawings, is ill-advised and can lead to erroneous values." The accuracy of a critical speed analysis based on measuring tire mark radius in a diagram would obviously depend on the accuracy of the measurement of the tire mark radius in the diagram. Tire marks located accurately with survey or scanning equipment or with photogrammetry would minimize error associated with the radius measurement. An accurate diagram also allows the analyst to verify that the vehicle was traveling an approximately circular path when the tire marks were deposited.

One common method of measuring the radius – applicable to measurements taken on a diagram or in the field – is to measure a chord and middle ordinate of either the vehicle center of gravity path or of one of the tire marks. These measurements are then converted to a radius measurement using the equation below. In this equation, c is the length of the chord and m is the length of the middle ordinate.

$$R = \frac{c^2}{8m} + \frac{m}{2} \tag{6.13}$$

Daily indicated that the radius measurement should be taken on the tire mark deposited by the front tire of the side of the vehicle that was leading the yaw. He also indicated that half of the vehicle track width should be subtracted from the measurement. Bartlett et al. [20] reported testing of investigators' ability to determine the radius of an arc by measuring the cord and middle ordinate. This testing was conducted at IPTM's 2001 Special Problems conference. The arc used for the testing was created using white chalk on black asphalt, and it had a constant radius of 182 ft (55.5 m). Eighteen investigators measured the arc, and as the authors note, nine of them chose a chord length of 50 ft (15.2 m). The longest chord utilized was 57.5 ft (17.5 m) and the shortest was 30 ft (9.1 m). The radius calculated by the participants based on their measured cords and middle ordinates ranged from 173.5 to 193.1 ft (52.9 to 59.7 m).

In a second test reported by the same authors, a yaw mark was created on a dry concrete surface using a sport utility vehicle. Participants measured the radius of the yaw mark, again by measuring a cord and middle ordinate and then calculating the radius. The authors noted that most participants used a 30-m (100 ft) tape to measure the chord and a 7.6-m (25 ft) tape to measure the middle ordinate. The authors noted that "two data points (one high and one low) were discarded because they were approximately 3 or more standard deviations from the mean. The standard deviation of the 30 remaining measurements was 4.6% of the mean." The calculated values of the radius were normally distributed.

In a 1997 article, Sledge [11] reviewed the literature and concluded that there was no consensus as to whether the critical speed formula should use the lateral or longitudinal friction capabilities of the tire and as to whether the peak or slide friction should be used. That said, many empirical studies have examined this issue and the most common method is to utilize the average, or steady-state deceleration rate, from a straight-line maximal braking test. This appears to be the consensus in 2017 based on the SAE Surface Vehicle Recommended Practice titled "Use of the Critical Speed Formula" [12].

Dickerson [21] drove an instrumented 1988 Ford Escort GT through a series of maneuvers (right circle turns, right step steers, and left/right double step steers) that produced yaw marks and used these yaw marks to test the critical speed formula. Testing was performed on a concrete surface. The curvature of the yaw marks was documented using electronic surveying equipment. Dickerson indicated that the radius of the left front tire mark was used in his calculations. He used the longitudinal slide friction coefficient in conducting his analysis. Dickerson's analysis resulted in calculated speeds that were always higher than the actual speed. He stated that the error increased with increasing slip angle and decreasing lateral acceleration. The lowest errors were obtained when the slip angle was below 25 degrees and the lateral acceleration was at or above the measured coefficient of friction. This stands to reason since the formula is derived assuming the lateral acceleration of the vehicle is exceeding the available friction coefficient and that the vehicle has not developed significant sideslip. One potential issue that Dickerson's results raise is the potential that a vehicle could deposit yaw marks at a lateral acceleration level below the available friction coefficient. Dickerson did not discuss this issue and does not present any photographs of the tire marks he used in his analysis.

Brach [22] used the critical speed formula to analyze a series of simulated yaw tests. He reported that the accuracy of the critical speed formula varied depending on where along the tire marks the radius was measured and whether the driver accelerated, coasted, or braked while the tire marks were deposited. Specifically, he stated that for the cases where the vehicle was coasting "using the early part of the trajectory to determine the radius R leads to a critical speed higher than the true value but all within 5%. Using the later part of the path to determine the radius R in all cases leads to a critical speed below the true value..." In the cases with braking, the calculated critical speed "is almost 20% low. In all cases, using the leading front tire (LFT) path to estimate R gives a critical speed value lower than when the center of gravity path is used. For early path estimates this improves the accuracy, but has the opposite effect for later path estimates." Runs where the vehicle was accelerating showed "the least sensitivity to which part of the trajectory is used to establish R." Brach also analyzed 94 yaw tests run by the California Highway Patrol. He reported that "for all driver control modes, the experimentally estimated speeds from the CSF are less than the measured speeds, with the highest accuracy for [acceleration] through [coasting] to the lowest for [braking]." Thus, Brach's study yielded a conclusion opposite that of Dickerson's.

Cliff [23] conducted twenty-two yaw tests on an asphalt surface using a 1991 Honda Accord EX-R traveling at speeds between 70 and 120 kph (43.5 and 74.6 mph). The antilock braking system of the test vehicle was bypassed, and a non-ABS was installed with two adjustable brake proportioning valves. Cliff reported that for half of the tests, "about 30% braking was applied." Using the critical speed formula, he concluded that the critical speed formula underestimated the actual speed by an average of 3.5 kph (2.2 mph) when the sliding coefficient of friction was used and overestimated, the actual speed by an average of 9.3 kph (5.8 mph) when the peak coefficient of friction was used. The coefficients of friction were measured using straight-line maximal braking tests.

Bartlett [24] reported a comparison of critical speed calculations to the measured speeds from yaw tests on dry pavement, dry gravel, dry grass, and a split grass/pavement surface. For conducting the critical speed calculations for each test, the vehicle's path radius was determined by measuring a chord and middle ordinate on the tire mark deposited by the front, outboard tire. The calculated radius was then reduced by one-half of the track width to approximate the center-of-gravity path radius. Chord lengths of 9-15 m (30-50 ft) were used. Bartlett reported nineteen tests conducted on dry pavement with initial vehicle speeds ranging from 60 to 100 kph (37.3 to 62.1 mph). Among these nineteen tests, and those reported for other surfaces, were tests that included braking of sufficient severity to activate the ABS and tests that included wide open throttle. Bartlett reported that "a conservative speed resulted in all cases when using the skidding friction value."

Thirteen tests were reported for a dry gravel surface. For these tests, Bartlett used the friction coefficient obtained from a brake-to-stop test with the ABS active. The critical speed results for these tests ranged from an underestimate of 10% to an overestimate of 4%. Bartlett reported four yaw tests on "mowed dry healthy grass which was 70-150 mm tall." Bartlett reported that "CSF results for the grass-only tests using the ABS-to-stop friction value were essentially spot-on accurate or slightly conservative." Finally, Bartlett reported five split-surface tests (grass/pavement). A weighted average coefficient of friction was used, with the friction coefficient for the outboard surface being given a 2/3 weighting, and the friction coefficient for the inboard surface being given a 1/3 weighting. The critical speed calculations for most of these tests "were very close or somewhat conservative."

Bartlett concluded that "the results here bolster earlier reports that the CSF model is tolerant of braking...as long as wheels are not locked, though some literature suggests that even that is not a problem. The CSF has [also] been shown to work well under full throttle application on dry pavement...The CSF model has been shown to work reasonably well on nominally flat unpaved gravel and grass surfaces." Lambourn [25] noted the reason that for vehicles with antilock brakes, the CSF could be tolerant of braking: "if a driver is steering so hard as to take up most of the available tyre road friction, the response of the anti-lock system to a simultaneous demand for heavy braking will be to keep the actual braking effort at a low level in order to prevent wheel locking. The situation should then be very similar to what happens when there is only light braking or no braking at all..." Based on his testing of the CSF, Lambourn concluded that the CSF "can still be usefully employed when a driver has braked hard enough to operate the ABS. It is possible that the error in the method is somewhat greater than in non-ABS vehicles, but if that is the case, the large errors are always of underestimation."

Asay reported results of critical speed analysis of the initial set of tire marks (from the leftward steering input) for 7 of the 8 tests he published in 2010. Asay obtained the radius for each test by first using a CAD diagram of the tire marks to reconstruct positions of the test vehicle at discrete points on the tire marks. He then used several of these vehicle positions to measure the radius of the arc followed by the vehicle's center of gravity. Positions were selected in the area where the peak lateral acceleration occurred. The friction coefficient for the roadway was assumed to be equal to the deceleration rate obtained during a straight-line maximal braking test with the ABS disabled. This testing utilized a Vericom 2000 performance measurement device and resulted in a friction coefficient of 0.772. For these 7 tests, the largest error in speed obtained with the critical speed formula was 1.1 mph (1.8 kph). For 3 of the tests, the critical speed formula overestimated the measured speed, and for the other 4, it underestimated the speed. Asay observed that application of the critical speed formula along sections of the tire marks not associated with the peak lateral acceleration "did not yield as favorable results. The critical [speed] formula applied to these other sections did not predict the measured speed adequately, presumably because the vehicle was not cornering at a critical speed even though cornering tire marks were made."

The SAE Surface Vehicle Recommended Practice titled "Use of the Critical Speed Formula" states that the "SAE Critical Speed Task Force...has determined the [critical speed] method to be valid within its limits of uncertainty when used correctly...Prior research has shown the critical speed method to be accurate to approximately −13.5% (erring with a calculated speed to low) to +10% (erring with a calculated speed too high)" [12].

Coefficient of Friction

The accuracy of the BNP/NCB model, the CRASH model, and the critical speed formula depends, in part, on the accuracy of the coefficient of friction selected by the reconstructionist. Typically, in a reconstruction, the actual value of the coefficient of friction is unknown, but a reasonable range of values can be chosen based on test data in the literature for the involved surface type. This section reviews a number of studies that aid the reconstructionist in selecting an appropriate range of values.

Dry Asphalt/Concrete – In an unpublished, but publically available study, Bartlett [26] summarized seventeen published and unpublished studies related to the available coefficient of friction for passenger vehicles on dry asphalt. He concluded, "Evaluating the reported friction values from multiple different sources using a variety of vehicles, drivers, surfaces, measuring tools, and analysis techniques gives average value in the 0.76g range with standard deviations of 0.04 to 0.07g. Given an average of 0.76 ± 0.06g, 95% of all cases can be expected to fall between 0.64g and 0.88g."

In a separate study based on a database of tests provided by Baker Materials Engineering, Bartlett [27] reported a similar range for locked-wheel braking deceleration rates for vehicles *without antilock brakes*. For maximal braking on dry pavement *with antilock brakes*, Bartlett [27] reported deceleration rates of 0.821 ± 0.067, an increase of about 8% due to the presence of antilock brakes. Lambourn reported average increases in maximal braking deceleration rates on dry asphalt of about 12% (0.1g) due to the presence of antilock brakes [25]. Lambourn tested 6 vehicles with initial speeds between 50 and 100 kph (37.3 and 62.1 mph).

Bartlett's study also noted "a two-tailed Student's t test of published data from skid tests on concrete and asphalt roadways showed that those two data sets are statistically indistinguishable." In his study, Bartlett also reported maximal braking tests he ran from approximately 40 mph (64.4 kph) with four different test vehicles – a standard sedan, a large SUV, a cargo van, and a high-performance vehicle. Each vehicle was tested with and without ABS on a concrete roadway and on a gravel roadway. Bartlett found that "on average, ABS-drag-factors on concrete were 15% higher than locked-wheel values. In contrast, the ABS-drag factors on gravel were an average of 16% lower."

Heinrichs [28] performed 252 skid-to-stop tests from initial speeds of 20, 40, 60, and 80 kph (12.4, 24.9, 37.3, and 49.7 mph) using three different radial tire grades (economy, touring, and performance) with a single 1991 Honda Accord EX-R on a single traveled asphalt road surface in "fair to good condition and without defects." The test vehicle was modified to remove the antilock braking system. The tires used on the test vehicle were initially new, but they were subjected to 600 km (373 miles) of combined city and highway driving prior to the testing. The coefficients of friction during skidding that were reported by Heinrichs were similar to those reported in the studies reviewed by Bartlett at speeds of 60 and 80 kph (37.3 and 49.7 mph). The coefficients of friction at speeds of 20 and 40 kph (12.4 and 25.9 mph) were higher. Heinrichs reported the following formula that captured the speed dependence of the coefficients of friction for skidding in his study:

$$\mu_{skid} = 0.7043 + 0.2453e^{-0.3124 s_{skid}} \tag{6.14}$$

In this formula, s_{skid} is the skid distance.

Leiss [29] reported tests with a 2012 BMW 328i sedan, on both dry and wet asphalt, using three different tire types - summer (high performance), all season, and winter. He conducted tests with and without ABS active and all of his tests had a nominal initial speed of 45 mph. Leiss reported a total of 72 tests. With ABS on dry asphalt, the *summer tire* produced an average deceleration of 0.98g (6 tests), with a range of 0.971 to 0.998g. Without ABS on dry asphalt, this same tire produced an average deceleration of 0.70g (6 tests) with a range of 0.677 to 0.727g. For these 6 tests, Leiss noted that "the deceleration data becomes noisy after 1.25 s during all of the runs due to the locked tire friction causing the rear suspension to enter into a resonant hop frequency." With ABS on wet asphalt, this summer tire produced an average deceleration rate of 0.95g (6 tests) with a range of 0.927 to 0.967g. Without ABS on wet asphalt, this tire produced an average deceleration rate of 0.68g (6 tests) with a range of 0.657 to 0.700g.

With ABS on dry asphalt, the *all-season tire* produced an average deceleration of 0.88g (6 tests), with a range of 0.868 to 0.889g. Without ABS on dry asphalt, this same tire produced an average deceleration of 0.73g (6 tests) with a range of 0.713 to 0.749g. With ABS on wet asphalt, this all-season tire produced an average deceleration rate of 0.85g (6 tests) with a range of 0.823 to 0.869g. Without ABS on wet asphalt, this tire produced an average deceleration rate of 0.61g (6 tests) with a range of 0.595 to 0.612g. With ABS on dry asphalt, the *winter tire* produced an average deceleration of 0.79g (6 tests), with a range of 0.767 to 0.798g. Without ABS on dry asphalt, this same tire produced an average deceleration of 0.83g (6 tests) with a range of 0.802 to 0.835g. With ABS on wet asphalt, this winter tire produced an average deceleration rate of 0.72g (6 tests) with a range of 0.698 to 0.770g. Without ABS on wet asphalt, this tire produced an average deceleration rate of 0.54g (6 tests) with a range of 0.488 to 0.580g.

Gravel – Based on skid test data provided by Baker Materials Engineering, Bartlett [27] reported longitudinal friction coefficients of 0.580 ± 0.074 for braking without ABS and of 0.483 ± 0.066 for braking with ABS. Based on his own testing, Bartlett reported an average locked-wheel deceleration rate of 0.623g (SD = 0.058g) and an average ABS deceleration rate of 0.495g (SD = 0.064g).

Case Study: Asay's Isuzu Rodeo Rollover Test (Test #2, 2010-01-0521)

This section continues with the analysis of the previous chapters of the Isuzu Rodeo rollover test that Asay reported in 2010. The roll phase of this test was analyzed in Chapters 3 and 4 and the trip phase in Chapter 5. This chapter analyzes the loss-of-control phase. In this test, the Isuzu was towed up to and released at a speed of 73.5 mph (118.3 kph). After release, the vehicle was steered with a sharp left steering input of approximately ¼ turn, causing the vehicle to travel to the left across the roadway and to yaw counterclockwise and develop a slip angle of sufficient magnitude that tire marks were deposited on the road surface. One second later, the first steering input was followed by a significant steering input back to the right. The vehicle continued off the left side of the road into the dirt, but reversed its yaw direction, developing a significant slip angle as it yawed in a clockwise manner. The vehicle deposited furrows in the dirt and then began rolling over.

Based on Asay's survey data, the tire marks and furrows deposited by the Isuzu during the loss-of-control and trip phases were diagrammed, and the vehicle positions depicted in Figure 6.9 were determined from this evidence. Analysis of the loss-of-control phase is influenced by how one chooses to model the trip phase, and any of the last five positions that are not included in the trip phase become a part of the loss-of-control phase. Speed

FIGURE 6.9 Loss-of-control and trip phase positions determined from the scene evidence.

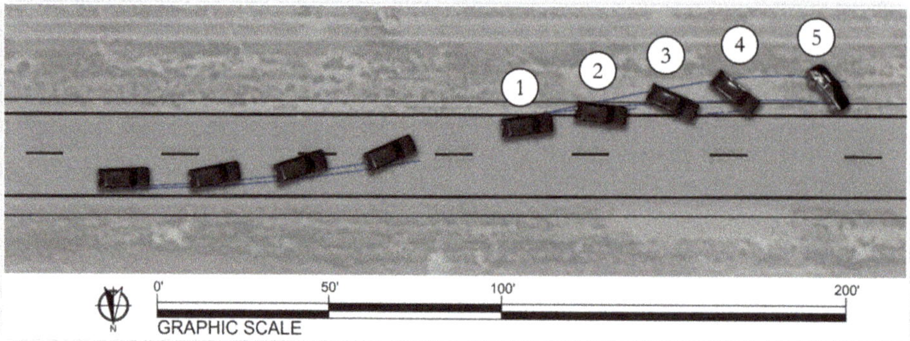

calculations for the loss of control were carried out based on the positions in Figure 6.9 for each of the trip phase definitions discussed in Chapter 5. These calculations, which are summarized in Table 6.2 (roll phase deceleration rate of 0.363g) and Table 6.3 (roll phase deceleration rate of 0.44g), used the CRASH model. For each position in Figure 6.9, the vehicle slip angle was measured and the distance between positions was measured.

TABLE 6.2 Summary of loss-of-control phase calculations (assuming roll phase deceleration rate of 0.363g)

Trip phase	Calculated speed at beginning of tire marks (mph), assuming no longitudinal force during trip	Calculated speed at beginning of tire marks (mph), matching speed at beginning of trip	Calculated speed at beginning of tire marks (mph), using actual longitudinal deceleration during trip
Positions 4-5	61.5 (−11.1)	67.2 (−5.4)	64.3 (−8.3)
Positions 3-5	62.1 (−10.5)	68.9 (−3.7)	66.4 (−6.2)
Positions 1-5	61.9 (−10.7)	71.5 (−1.1)	69.7 (−2.9)

TABLE 6.3 Summary of loss-of-control phase calculations (assuming roll phase deceleration rate of 0.44g)

Trip phase	Calculated speed at beginning of tire marks (mph), assuming no longitudinal force during trip	Calculated speed at beginning of tire marks (mph), matching speed at beginning of trip	Calculated speed at beginning of tire marks (mph), using actual longitudinal deceleration during trip
Positions 4-5	65.2 (−7.4)	67.3 (−5.3)	67.9 (−4.7)
Positions 3-5	65.7 (−6.9)	68.9 (−3.7)	69.8 (−2.8)
Positions 1-5	65.4 (−7.2)	71.5 (−1.1)	72.8 (+0.2)

FIGURE 6.10 Distance and slip angle measurements.

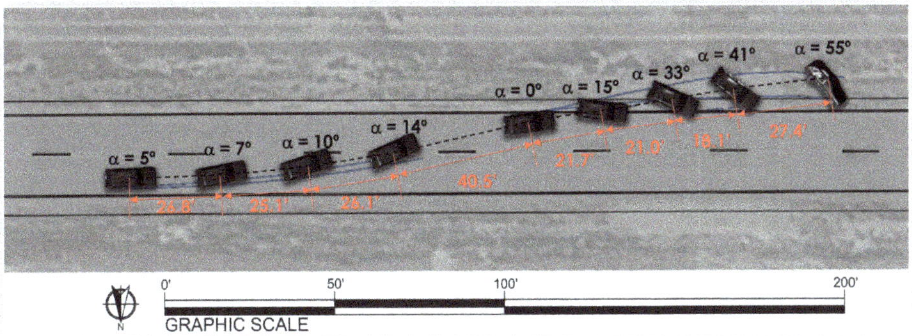

Figure 6.10 shows these distance and slip angle measurements. In these calculations, the slip angles for the front tires were not adjusted based on the steering inputs. Asay reported the vehicle was in neutral during this test, and so, no braking was applied to the vehicle during the loss of control. In addition, Asay reported a roadway coefficient of friction of 0.772. Based on the VBOX data, the actual speed at the beginning of the tire marks was 72.6 mph.

The speeds reported in the second columns of Tables 6.2 and 6.3 assume that there were no longitudinal forces present during the trip phase. Thus, these speeds carry the error from the trip phase into the analysis of the loss-of-control phase. The results in the third columns of Tables 6.2 and 6.3 assume as much longitudinal force during the trip phase as necessary to match the actual speed at the beginning of trip. Thus, these calculations test only the loss-of-control model for the different trip phase definitions. The results in the fourth columns of Tables 6.2 and 6.3 assume the actual longitudinal force for each trip phase duration as reported in Chapter 5. The results of Table 6.2 suggest, first, that if the longitudinal forces from steering and furrowing during the trip phase are not incorporated, the speed calculated speeds will likely be underestimated. A conservative speed estimate can be obtained without including these forces.

References

1. Carter, N., Beauchamp, G., and Rose, N., "Comparison of Calculated Speeds for a Yawing and Braking Vehicle to Full-Scale Vehicle Tests," SAE Technical Paper 2012-01-0620, 2012, doi:10.4271/2012-01-0620.

2. Martinez, J. and Schlueter, R., "A Primer on the Reconstruction and Presentation of Rollover Accidents," SAE Technical Paper 960647, 1996, doi:10.4271/960647.

3. Bakker, E., Nyborg, L., and Pacejka, H., "Tyre Modelling for Use in Vehicle Dynamics Studies," SAE Technical Paper 870421, 1987, doi:10.4271/870421.

4. Brach, R. and Brach, R., "Modeling Combined Braking and Steering Tire Forces," SAE Technical Paper 2000-01-0357, 2000, doi:10.4271/2000-01-0357.

5. Brach, R.M. and Brach, R.M., *Vehicle Accident Analysis and Reconstruction Methods*, (SAE International, 2005), ISBN: 0768007763.

6. Brach, R.M. and Brach, R.M., "Tire Models Used in Accident Reconstruction Vehicle Motion Simulation," Presented at *XVII Europäischen für Unfallforschung und Unfallanalyse (EVU)-Conference*, Nice, France, 2008.

7. Brach, R. and Brach, R., "Tire Models for Vehicle Dynamic Simulation and Accident Reconstruction," SAE Technical Paper 2009-01-0102, 2009, doi:10.4271/2009-01-0102.

8. Beauchamp, G., Hessel, D., Rose, N., Fenton, S. et al., "Determining Vehicle Steering and Braking from Yaw Mark Striations," *SAE Int. J. Passeng. Cars - Mech. Syst.* 2, no. 1 (2009): 291-307, doi:10.4271/2009-01-0092.
9. Cliff, W. and Bowler, J., "The Measured Rolling Resistance of Vehicles for Accident Reconstruction," SAE Technical Paper 980368, 1998, doi:10.4271/980368.
10. Sledge, N. and Marshek, K., "Formulas for Estimating Vehicle Critical Speed from Yaw Marks-A Review," SAE Technical Paper 971147, 1997, doi:10.4271/971147.
11. Sledge, N. and Marshek, K., "Vehicle Critical Speed Formula-Values for the Coefficient of Friction-A Review," SAE Technical Paper 971148, 1997, doi:10.4271/971148.
12. SAE Surface Vehicle Recommended Practice, "Use of the Critical Speed Formula," SAE J2969, January 2017.
13. Daily, J., Shigemura, N., and Daily, J., *Fundamentals of Traffic Crash Reconstruction*, Volume 2 of the Traffic Crash Reconstruction Series, (Jacksonville, FL, Institute of Police Technology and Management), ISBN 1884566634.
14. Fricke, L.B., *Traffic Crash Reconstruction*, 2nd ed., (Evanston, IL, Northwestern University Center for Public Safety, 2010), ISBN 0912642300.
15. Richardson, S., Orton, T., Josevski, N., and Jones, C., "A Critique of Critical Speed Yaw Mark Research," SAE Technical Paper 2012-01-0600, 2012, doi:10.4271/2012-01-0600.
16. Masory, O., Delmas, S., Wright, B., and Bartlett, W., "Validation of the Circular Trajectory Assumption in Critical Speed," SAE Technical Paper 2005-01-1189, 2005, doi:10.4271/2005-01-1189.
17. Baumann, F.W., Schreier, H., and Simmermacher, D., "Tire Mark Analysis of Modern Passenger Vehicle with respect to Tire Variation, Tire Pressure and Chassis Control System," *SAE Int. J. Passeng. Cars - Mech. Syst.* 2, no. 1 (2009): 333-344, doi:10.4271/2009-01-0100.
18. Brown, J.C., Grimm, M., and Hansen, D., "Relationship between Anti-Lock Tire Mark Length and Speed Change," SAE Technical Paper 2007-01-0723, 2007, doi:10.4271/2007-01-0723.
19. Seipel, G., Winner, H., Baumann, F., and Hermanutz, R., "Approach to Determine Slip Values Based on the Intensity of Tire Marks with respect to Tire and Road Properties," SAE Technical Paper 2013-01-0781, 2013, doi:10.4271/2013-01-0781.
20. Bartlett, W., Wright, W., Masory, O., Brach, R. et al., "Evaluating the Uncertainty in Various Measurement Tasks Common to Accident Reconstruction," SAE Technical Paper 2002-01-0546, 2002, doi:10.4271/2002-01-0546.
21. Dickerson, C., Arndt, M., Arndt, S., and Mowry, G., "Evaluation of Vehicle Velocity Predictions Using the Critical Speed Formula," SAE Technical Paper 950137, 1995, doi:10.4271/950137.
22. Brach, R., "An Analytical Assessment of the Critical Speed Formula," SAE Technical Paper 970957, 1997, doi:10.4271/970957.
23. Cliff, W., Lawrence, J., Heinrichs, B., and Fricker, T., "Yaw Testing of an Instrumented Vehicle with and without Braking," SAE Technical Paper 2004-01-1187, 2004, doi:10.4271/2004-01-1187.
24. Bartlett, W. and Wright, W., "Yaw Tests on Pavement, Gravel, and Grass with ESC and ABS Braking," *1st Joint ITAI-EVU Conference*, Hinckley, UK, 2009.
25. Lambourn, R.F., "Braking and Cornering Effects with and without Anti-Lock Brakes," SAE Technical Paper 940723, 1994, doi:10.4271/940723.
26. Bartlett, W., "Skidding Friction: A Review of Recent Research," unpublished, available at http://www.mfes.com/friction.html.
27. Bartlett, W. and Wright, W., "Braking on Dry Pavement and Gravel with and without ABS," SAE Technical Paper 2010-01-0066, 2010, doi:10.4271/2010-01-0066.
28. Heinrichs, B.E., Allin, B., Bowler, J., and Siegmund, G., "Vehicle Speed Affects Both Pre-Skid Braking Kinematics and Average Tire/Roadway Friction," *Accident Analysis and Prevention* 36 (2004): 829-840, doi:10.1016/j.aap.2003.08.002.
29. Leiss, P.J., Steven, B., and Gary, D., "Tire Friction Comparison of Three Tire Types," SAE Technical Paper 2013-01-0783, 2013, doi:10.4271/2013-01-0783.

Appendix

FIGURE 6.11 Speed analysis results for Test #1, SI units.

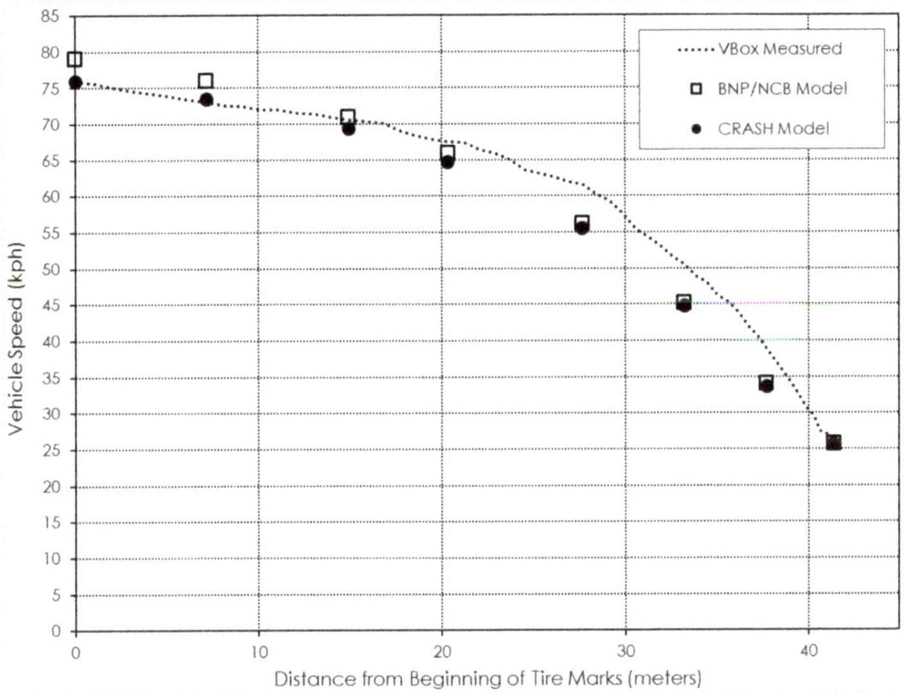

FIGURE 6.12 Speed analysis results for Test #2, SI units.

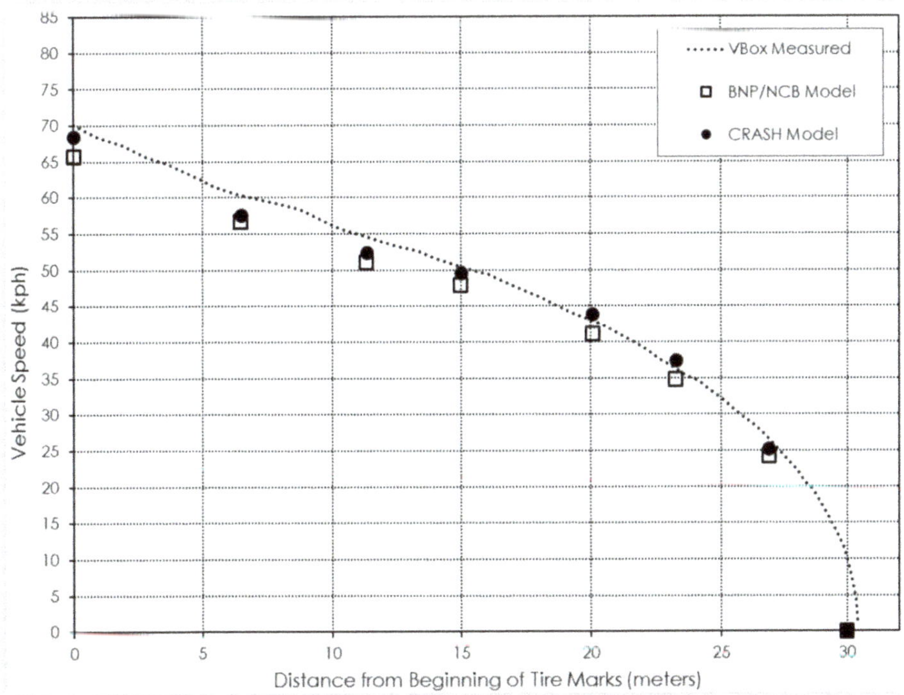

CHAPTER 7

Analyzing Tire Mark Striations

When a vehicle yaws, its velocity direction deviates from its heading and its tires slip sideways. Once a sufficient lateral slip level is reached, the tires deposit tire marks, and sometimes these tire marks contain striations. As an example, Figure 7.1 depicts a set of striated tire marks. Driver steering and braking inputs affect the characteristics of yaw mark striations. The analysis techniques discussed in this chapter allow for these inputs to be quantified from the striation evidence, in some instances.

Mechanisms That Create Striation Evidence

The lateral sliding of the tire during a yaw deforms the tire carcass, loading the leading shoulder of the tire (the shoulder is the area where the tread or bottom of the tire and the tire sidewall meet). Figure 7.2 depicts lateral deformation of a tire during a yaw maneuver. Figure 7.3 depicts the wear to the shoulder blocks of a tire as a result of yaw testing.

From the wear evidence on the tire shown in Figure 7.3, the shoulder blocks are clearly involved in creating the tire marks. Nevertheless, two theories have been proposed to explain how striations are deposited: in-plane buckling of radial tires [1, 2, 3, 4]) and

FIGURE 7.1 An example of striated tire marks.

FIGURE 7.2 Lateral tire deformation during a yaw.

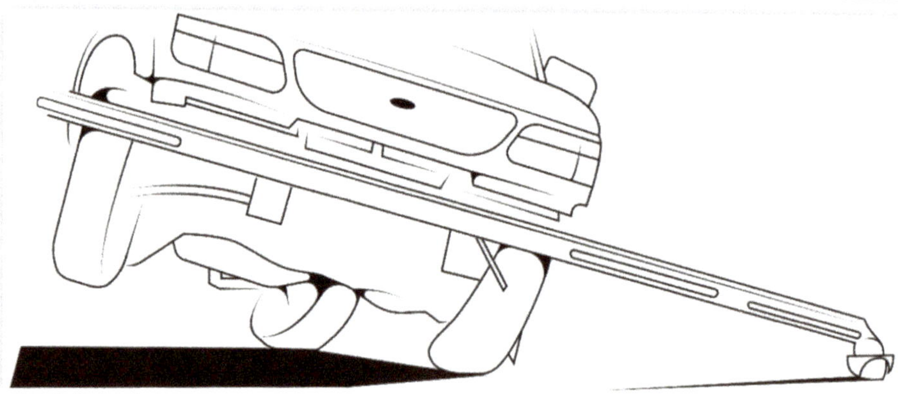

point loading of the shoulder tread blocks [3, 5, 6, 7, 8, 9, 10]. It was shown that in-plane buckling can occur in the contact patch, especially when the tire has no tread pattern [2, 4] or the tire pressure has been reduced to very low values [3]. The results by Yamazaki suggest that bias ply tires do not deposit striations because bias ply tires are not susceptible to in-plane buckling [2, 4]. However, Reveley successfully deposited striated tire marks using a bias ply tires [10]. Reveley's results are consistent with the tread shoulder blocks being involved in the creation of striated marks. Gardner's testing with large tread block tires also created striated marks. However, the testing with highway tires without a defined tread block pattern did not create striation marks (unless the tire pressure was reduced substantially) [3].

In general, testing involving tires with an aggressive shoulder block pattern deposited clear striations, whereas tires without an aggressive shoulder block did not (unless the tire pressure was reduced substantially). A mathematical relationship between the striation marks and the shoulder tread blocks was also reported by Beauchamp, further suggesting that the shoulder tread blocks made the marks [8]. The shoulder tread blocks

FIGURE 7.3 Wear to the shoulder blocks of a tire from yaw testing.

from Beauchamp's testing also exhibited heavy wear, again consistent with the tread blocks depositing the striation evidence (Figure 7.3).

Some research does suggest that in-plane buckling can occur, but it is more likely to be the source of striations if deposited by a tire with no tread pattern or an underinflated tire. Tires with more aggressive shoulder tread block patterns are more likely to deposit striated tire marks when properly inflated. Examination of wear to the tire can reveal whether in-plane buckling or the tread blocks deposited the marks. Figure 7.4 depicts wavy wear to the tire as a result of radial tire in-plane buckling from testing performed by [2].

Identifying which mechanism created striations can be important, or not, depending on the analysis technique employed. In 2009, Beauchamp [8] introduced two methods for analyzing striations. The first, called the longitudinal slip equation, is used to determine the braking level. The longitudinal slip equation considers the slip angle of the tire and angle of the striation. This equation is based on the striation direction being an indication of the overall tire force direction. This is true regardless of the mechanism that led to the striations being deposited. Therefore, analysis of striations with the longitudinal slip equation is valid, independent of the mechanism that creates the marks. The second analysis method uses the slip angle equation to determine the steering input. The slip angle equation also considers the spacing between the striations and assumes that the striations are created by the tire shoulder blocks.

FIGURE 7.4 Tire wear as a result of in-plane buckling.

What Can Striations Tell Us?

Striations in tire marks have long been recognized as an indication of the level of braking used by a driver, and discussions of tire mark striations appear in various places in the accident reconstruction literature. For example, Fricke [11] discussed tire mark striations, stating that when a vehicle is yawing and no braking is applied to the vehicle, the striations are deposited parallel to the axle of the tire that deposited the marks. Fricke further indicated that adding increasing levels of braking would result in the striations moving more into line with the tire marks themselves. Fricke noted that acceleration, on the other hand, would angle the striations in the opposite direction as braking, typically making them more perpendicular to the tire mark. Fricke stated that steering would have no effect on the angle of the striations.

As another example, Reveley [10] reported a comparison of skid marks and yaw marks based on straight line braking tests and yaw tests with both radial and bias ply tires. Reveley noted that during the braking tests, striation marks were deposited that were parallel to the tire marks. Striation marks deposited by a yawing vehicle were described as being lateral to the tire mark direction. As a vehicle yawed past a slip angle of 90°, striations within the tire marks were transitioned from "lateral to parallel and back to lateral." In these tests, steering did affect the orientation of the striations. Similarly, Reveley noted that braking during yawing affected the striation orientation. These observations led to the conclusion that striations parallel to a tire mark are indicative of braking and striations lateral to the tire mark are indicative of no braking.

Similarly, Hague ran numerous vehicle tests and published two studies that discussed striations in tire marks [12, 13]. Tests were run with the vehicles being accelerated, braked, or just coasting. Hague examined the relationship between the angle of the striations and the driver acceleration or braking inputs. In general, Hague concluded that braking during yawing had the effect of bringing the striations more in line with the tire mark than they were without braking. Acceleration had a similar effect, but in the opposite direction. In Hague's second study, the test vehicles were instrumented with longitudinal and lateral accelerometers so that the angle of striations could be compared to the net tire force acting on the vehicle. It was hypothesized that the net tire force on the vehicle is equal to the weighted average, by vertical load, of the striation angles of the four tires. Ultimately, Hague was not able to prove this with these tests but nonetheless maintained that the angle of striations were an indication of the angle of the net tire force.

Models for Analyzing Striations

Contrary to what Fricke stated, for a vehicle that is yawing without any braking by the driver, the striations are deposited perpendicular to the tire. However, while a vehicle is yawing, the driver may apply the brakes. The level of braking affects the longitudinal slip of the tires and the orientation of the striations in the tire marks. Specifically, the striation and force direction is perpendicular to the tire in the absence of braking and transitions towards the direction of the tire mark as the brakes are applied. The trend is depicted graphically in Figure 7.5. The tires in Figure 7.5 are moving up the page as indicated by the blue arrow.

Beauchamp [8] hypothesized that "the striations indicate the [travel] path of individual portions of the tire. Since frictional forces act to oppose the direction of motion... these striations are also indicative of the direction of the frictional force acting on these portions of the tire." Given this, any factor that affects the resultant direction of the tire force also affects the striation direction and, potentially, the striation spacing. This means

FIGURE 7.5 The effect of braking on striation orientation.

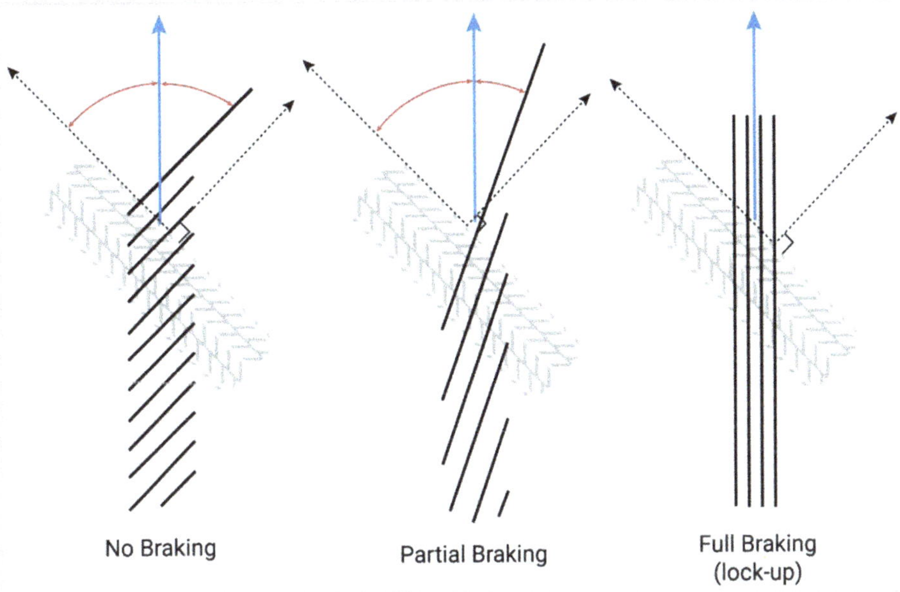

that acceleration, braking, and steering all influence the orientation and spacing of the striations. Beauchamp developed equations that related the angle and spacing of striations, along with the tire slip angle, to the longitudinal and lateral slip levels of the tire that deposited the mark, providing "a link between physical evidence deposited on a roadway during a crash (the tire mark striations) and actions taken by the driver during that crash (steering and braking inputs)."

Slip Angle from Striations (Steering)

Figure 7.6 is a diagram depicting a tire (assumed to be rigid) depositing a striated yaw mark. The coordinate system depicted in this figure is right handed, with the z-axis pointing into the page and the x-axis aligned with the tire's heading. The tire slip angle is identified with the symbol α. The striation angle relative to the velocity direction is identified with the symbol θ, and the striation angle relative to the tire heading is identified with the symbol Ψ. These angles have a positive sign when they are measured consistent with how they are depicted in this figure. The contact patch of the tire is identified with a gray box and the length of the contact patch is designated *CPL*. The striations depicted in this figure are neither perpendicular to the tire heading nor parallel to the velocity direction, indicating that braking is occurring but the wheel is not locked. The spacing between striations, measured along the velocity direction, is designated with the symbol D_S and measured on the tire with the symbol D_T.

In practice, the angle θ and the distance D_S can be measured on a striated tire mark. The tread block spacing, D_T, can be measured on the tire. Given the geometry depicted in Figure 7.6, Equation (7.1) yields Ψ, given θ, D_S, and D_T:

$$\Psi = \sin^{-1}\left(\frac{D_S \cdot \sin\theta}{D_T}\right) \tag{7.1}$$

FIGURE 7.6 Yawing tire depositing striations.

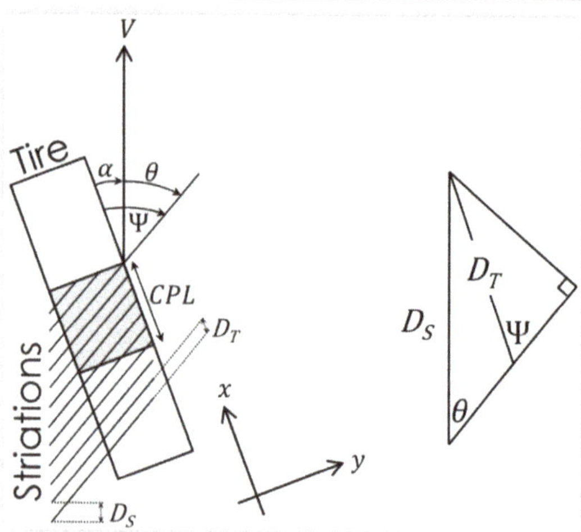

FIGURE 7.7 The motion of a tire while It deposits a striation.

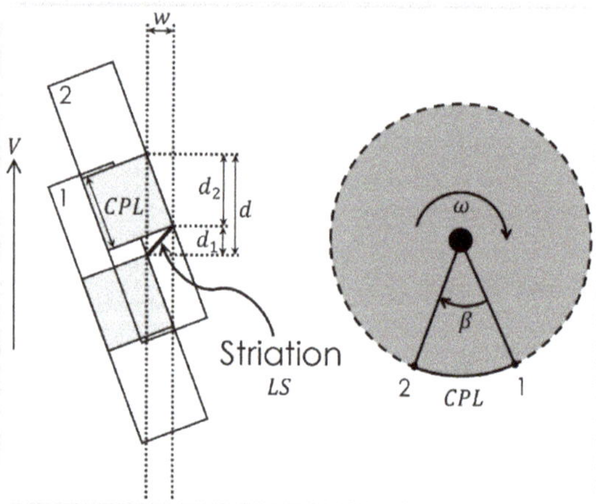

Given the geometric relationships between Ψ, α, and θ, Equation (7.2) gives an expression for the slip angle. When multiple tire marks are deposited on the roadway, the vehicle yaw angle at a particular point can be determined by aligning the vehicle with its tire marks. Using Equation (7.2), the slip angle of each tire can also be determined. Rear tire slip angles are equal to the slip angle of the vehicle in the absence of tire misalignment or toe in/out design differences. The difference between the heading direction of the front tires and the heading direction of the vehicle is the steering angle at the front tires:

$$\alpha = \sin^{-1}\left(\frac{D_S \cdot \sin\theta}{D_T}\right) - \theta \tag{7.2}$$

Longitudinal Slip from Striations (Braking)

Figure 7.7 depicts the motion of a tire while it deposits a single striation. While the tire deposits the striation mark depicted by the bold line, it moves from Position 1 to Position 2. In Position 1, a point on the circumference of the tire marked with a black dot has just come into contact with the ground at the front of the contact patch. While the striation is being deposited, the tire rotates and that point on the tire moves towards the rear of the contact patch before lifting away from the ground at Position 2. The image on the right of this figure depicts a side view of the tire and the circumference of the tire that is traversed as the tire rolls from Position 1 to 2 (*CPL*). Between Positions 1 and 2, the tire rotates through an angle of β radians. While the tire travels from Position 1 to 2, it deposits a striation of length *LS*. The portion of the tire mark deposited by the outboard shoulder of the tire has a width *w*.

Along with the orientation of the tire, the braking level (or longitudinal slip percentage) can also be calculated using some of the dimensions in Figure 7.7. Equation (7.3) gives the definition of longitudinal slip (*S*):

$$S = \frac{V \cdot \cos\alpha - \omega r}{V \cdot \cos\alpha} \tag{7.3}$$

In Equation (7.3), *V* is the velocity of the wheel hub, ω is the rotational speed of the tire, and *r* is the radius of the tire. When the tire is free rolling, this equation has a value of 0. When the rotation of the tire ceases entirely, this equation has a value of 1.

Given the geometry of Figure 7.7, Equation (7.3) can now be rewritten as follows:

$$S = \frac{\frac{d}{\Delta t}\cos\alpha - \frac{\beta}{\Delta t}r}{\frac{d}{\Delta t}\cos\alpha} \tag{7.4}$$

In this equation, Δt is the time that elapses while one striation is deposited. This time is the same in each of the terms of Equation (7.4) and so it cancels out, leaving the following equation:

$$S = 1 - \frac{\beta r}{d \cdot \cos\alpha} \tag{7.5}$$

The width of the portion of the tire mark deposited by the shoulder of the tire can be calculated in two ways with the following equation:

$$w = CPL \cdot \sin\alpha = LS \cdot \sin\theta \tag{7.6}$$

Equation (7.6) can then be rearranged to yield the contact patch length (*CPL*), as follows:

$$CPL = \frac{LS \cdot \sin\theta}{\sin\alpha} \tag{7.7}$$

The contact path length is also equal to βr. Now, consider the distances d, d_1, and d_2. These distances can be determined geometrically as follows:

$$d_1 = LS \cdot \cos\theta \tag{7.8}$$

$$d_2 = CPL \cdot \cos\theta = \frac{LS \cdot \sin\theta}{\sin\alpha} \cdot \cos\theta \tag{7.9}$$

$$d = d_1 + d_2 = LS \cdot \left(\cos\theta + \frac{\sin\theta}{\sin\alpha} \cdot \cos\theta \right) \tag{7.10}$$

Equations (7.7) and (7.10) can then be substituted into Equation (7.5), which after some algebraic manipulation yields the following:

$$S = 1 - \frac{\sin\theta}{\cos\theta\left(\cos\theta \sin\alpha + \sin\theta \cos\alpha\right)} \tag{7.11}$$

Using a trigonometric identity and the geometry of Figure 7.7, Equation (7.11) reduces to the following:

$$S = 1 - \frac{\sin\theta}{\cos\alpha \sin\left(\theta + \alpha\right)} = 1 - \frac{\sin\left(\Psi - \alpha\right)}{\cos\alpha \sin\Psi} \tag{7.12}$$

Applying a series of additional trigonometric identities, Equation (7.12) can be written in the following form:

$$S = \frac{\tan}{\tan\left(\theta + \alpha\right)} \tag{7.13}$$

Equation (7.13) can be multiplied by 100 to yield the slip percentage. This equation demonstrates that the longitudinal slip of a tire can be calculated if the angle of the striations can be measured and if the slip angle of the tire is known.

Equation (7.13) can also be derived from tire force models [14]. Figure 7.8 depicts a tire in the midst of a yaw while braking. The tire is moving up the page as indicated by the blue arrow. In Figure 7.8, the longitudinal and lateral forces are indicated with Q_x and Q_y, respectively. The total force, Q, which acts at an angle δ from the tire heading,

FIGURE 7.8 Tire in yaw with brakes applied.

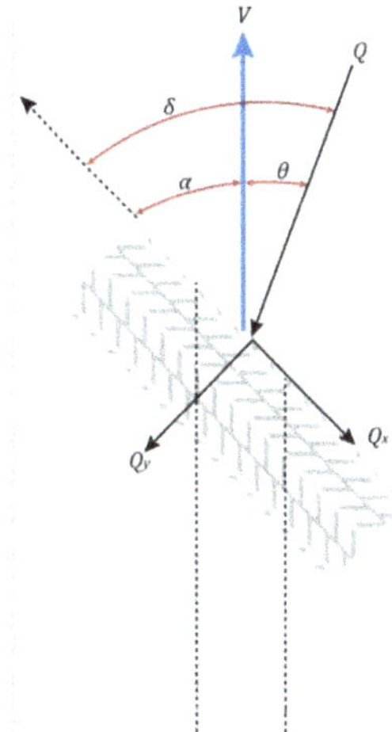

is also indicated. The angle α is the tire's slip angle, and θ is the angular difference between the tire hub's velocity direction and the force direction. Therefore, δ is equal to the sum of α and θ.

Assume now that the tire force is aligned with the striation that was deposited by that yawing tire. The angle θ between the velocity direction and the force direction would also be the striation angle. Using Figure 7.8, the angle θ and the forces Q_x and Q_y are related as follows:

$$\delta = \alpha + \theta = \tan^{-1}\left(\frac{Q_y}{Q_x}\right) \tag{7.14}$$

The Nicolas-Comstock-Brach expressions for Q_x and Q_y are then substituted, and the equation can be solved for longitudinal slip, s, as demonstrated in the following equations:

$$\alpha + \theta = \tan^{-1}\left(\frac{\dfrac{F_x(s)F_y(\alpha)\tan\alpha}{\sqrt{s^2\left(F_y(\alpha)\right)^2 + \left(F_x(s)\right)^2\left(\tan\alpha\right)^2}}}{\dfrac{F_x(s)F_y(\alpha)s}{\sqrt{s^2\left(F_y(\alpha)\right)^2 + \left(F_x(s)\right)^2\left(\tan\alpha\right)^2}}}\right) \tag{7.15}$$

$$\alpha + \theta = \tan^{-1}\left(\frac{\tan\alpha}{s}\right) \tag{7.16}$$

$$\tan(\theta + \alpha) = \frac{\tan\alpha}{s} \tag{7.17}$$

$$s = \frac{\tan\alpha}{\tan(\alpha + \theta)} \tag{7.18}$$

Equation (7.18) is equivalent to the equation derived previously from the classic definition of longitudinal slip - Equation (7.13).

Tire Mark Striations - Sensitivity and Uncertainty Analysis

This section explains how uncertainties in measurements of the striation orientation propagate through striation analysis with Equation (7.13) [9]. First, consider Figure 7.9 in which a new variable, κ, is defined. This variable represents the angle between a normal to the tire heading and the striation direction. Equation (7.19) is a reformulation of Equation (7.13) using κ instead of θ:

$$S = \frac{\tan\alpha}{\tan(90 - \kappa)} \tag{7.19}$$

Figure 7.9 shows how application of the brakes by the driver changes the orientation of the striations relative to the tire mark. When there is no braking, the striations are perpendicular to the tire heading and the angle κ is equal to zero. When the braking is of sufficient severity to stop the rotation of the tire, the striations become parallel to the wheel hub velocity and parallel to the tire mark. Figure 7.9 shows the case of no braking,

FIGURE 7.9 Transition in striation direction as a result of braking.

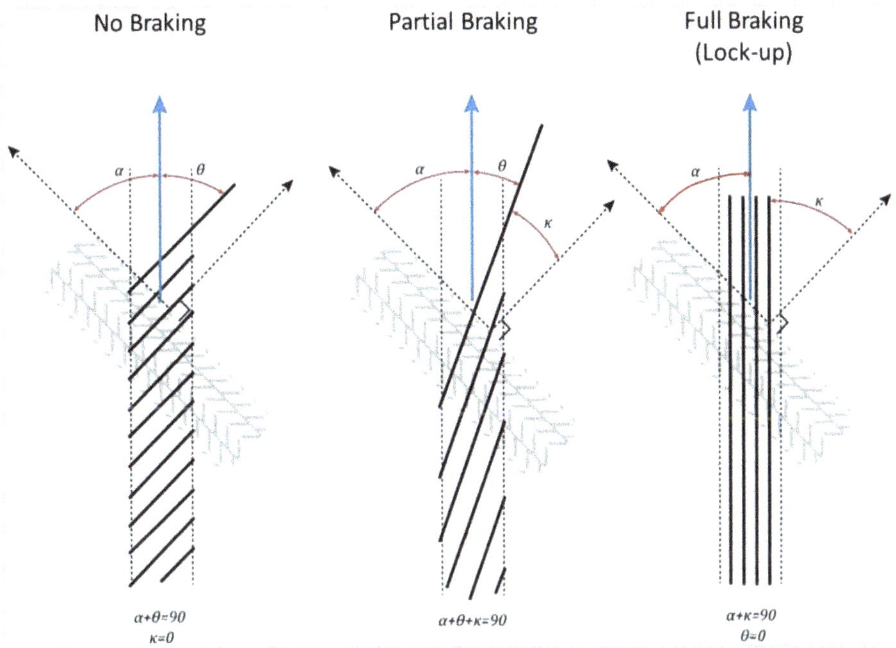

partial braking, and full braking. In all three cases, the tire is at the same slip angle, α. The velocity of the tires is towards the top of the page, as indicated by the blue arrows. In the case of no braking, $\alpha + \theta = 90^\circ$ and $\kappa = 0^\circ$. In the case of partial braking, $\alpha + \theta + \kappa = 90^\circ$. In the case of full braking, $\theta = 0^\circ$ and $\alpha + \kappa = 90^\circ$.

Now consider the effect of the slip angle. When the slip angle is small, there is a large angular difference between the value of κ associated with no braking and the value of κ associated with full braking. When the slip angle is large, there is a small difference between the value of κ associated with no braking and the value of κ associated with full braking. This means that striation analysis will become more sensitive to uncertainties in the striation orientation as the slip angle increases. This sensitivity is illustrated in Figure 7.10. On this graph, the longitudinal slip percentage is plotted as a function of the angle κ. Nine different curves are included on the graph, each representing a different slip angle (5°-85°, in 10° increments).

FIGURE 7.10 Slip percentage as a function of κ for several tire slip angles.

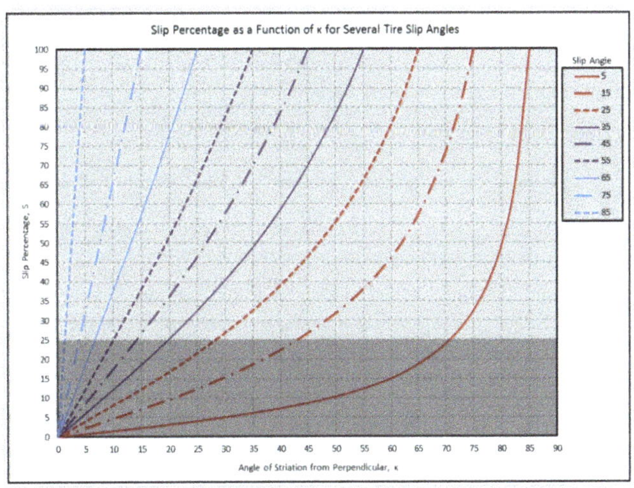

Maximum deceleration occurs at around 25% longitudinal slip for a typical passenger tire. The region where this buildup to maximum deceleration occurs is shaded with a darker gray in Figure 7.10. This graph shows that, at a slip angle of 5°, the striations transition through an angular change of 70° between the no-braking and full-braking deceleration scenarios. This means that for a low slip angle like 5°, the no-braking and heavy-braking cases are easily distinguishable. At an 85° slip angle, on the other hand, less than 2° separate the no-braking and full-braking cases. With this level of lateral slip, the reconstructionist is unlikely to be able to distinguish the two cases.

FIGURE 7.11 The variables of Equations (7.13) and (7.20).

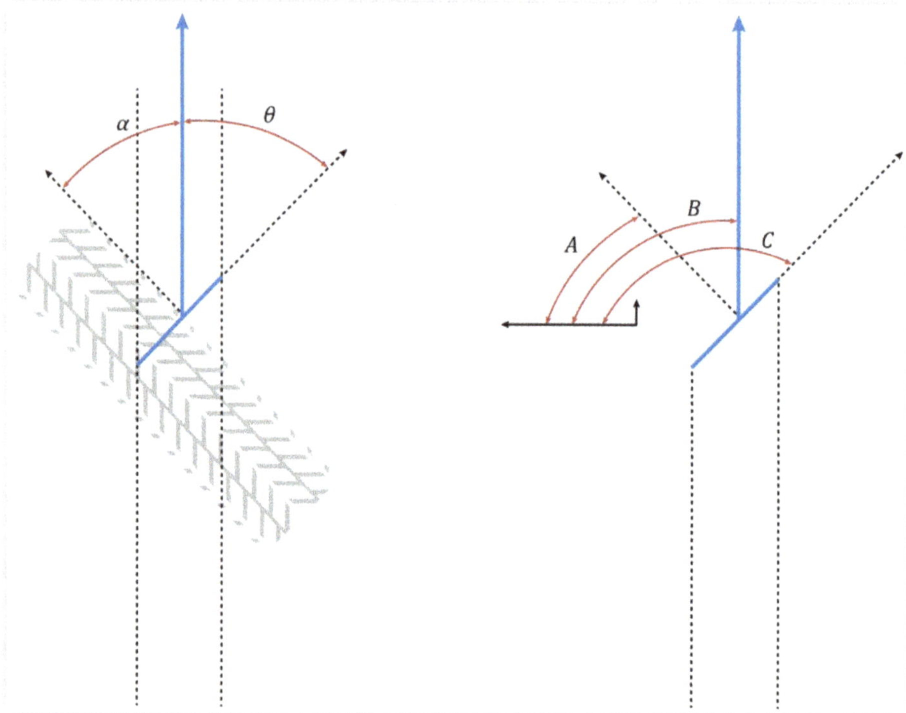

Uncertainties in α and θ are correlated since they are both measured in relation to the velocity direction (see the left side of Figure 7.11). Further, the uncertainties in α and θ have negative covariance, since an overestimation of α will likely be accompanied by an underestimation in θ, and vice versa (Taylor, 1982). In order to account for this, the Equation (7.13) can be rewritten in a fixed coordinate system. On the right side of Figure 7.11, the angles α and θ are redefined with three new angles, A, B, and C. With these three new angles, Equation (7.13) can be rewritten as Equation (7.20):

$$S = \frac{\tan(B-A)}{\tan(C-A)} \tag{7.20}$$

Equation (7.20) now contains three independent measurements. By rewriting Equation (7.13) in this way, an uncertainty analysis can be conducted that accounts for the negative covariance between the angles α and θ. Equation (7.21) yields the uncertainty in the longitudinal slip given the uncertainties in the three angles in these three angles δA, δB, and δC (Taylor, 1982):

$$\delta S = \sqrt{\left(\frac{\partial S}{\partial A}\delta A\right)^2 + \left(\frac{\partial S}{\partial B}\delta B\right)^2 + \left(\frac{\partial S}{\partial C}\delta C\right)^2} \tag{7.21}$$

The partial derivatives in Equation (7.21) are defined as follows:

$$\frac{\partial S}{\partial A} = \frac{\sec^2(C-A)\tan(B-A) - \sec^2(B-A)\tan(C-A)}{\tan^2(C-A)} \tag{7.22}$$

$$\frac{\partial S}{\partial B} = \frac{\sec^2(B-A)}{\tan(C-A)} \tag{7.23}$$

$$\frac{\partial S}{\partial C} = -\frac{\tan(B-A)\sec^2(C-A)}{\tan^2(C-A)} \tag{7.24}$$

Using Equations (7.21) through (7.24), the uncertainty in the longitudinal slip determined from the striation direction can be calculated. Beauchamp [9] conducted analysis with these equations and made the following observations:

1. Different combinations of slip angle and braking result in different levels of uncertainty in calculating longitudinal slip percentage from striations.
2. If the striations are perpendicular or nearly perpendicular to the tire ($\kappa \sim 0$), then the uncertainty in slip percentage increases as slip angle increases.
3. As the brakes are applied and longitudinal slip is generated, the uncertainty in slip increases at lower slip angles and decreased at higher slip angles. The transition occurred near a slip angle of 55°.
4. In general, the uncertainty in slip was greatest near wheel lock up (near 100% slip). However, maximum braking is achieved at a slip angle of approximately 25%. Between 25% and 75% slip, there is minimal change in how the vehicle is actually decelerating. Therefore, greater uncertainty at higher slip percentages is of little consequence to a speed analysis calculation.
5. Uncertainty in the longitudinal slip calculation was relatively large at the maximum slip angle studied here ($\alpha = 85^\circ$). It is unlikely that striation analysis would yield dependable results at this slip angle. Calculations of slip percentage from striations at high slip angles may yield meaningless results due to the uncertainties in measurement. However, at this slip angle, the vehicle is nearly sliding sideways, and application of the brakes does not significantly change the deceleration rate. So, even though striation analysis is most sensitive at higher slip angles, this is the region where the results of the striation analysis matter the least.

Case Study - Striation Analysis

In this section, a real-world crash is analyzed to demonstrate the application of striation analysis. In this crash, an SUV departed the left side of the road in a counterclockwise yaw. Once it exited the road, the vehicle tripped and then rolled over multiple times before coming to rest on its roof in the center median. The vehicle deposited striated tire marks while it was yawing on the roadway. The initial speed of the vehicle was in question. Figure 7.12 depicts the tire marks on the roadway and vehicle at rest in the median. The striations in the tire marks are visible, most notably, where the tire passed over the yellow lane line in the lower right corner of the figure.

FIGURE 7.12 Tire marks and rest position of the vehicle in the case study.

Prior to analyzing the tire marks, a crash scene diagram and vehicle model were created. This process involved the following steps:

1. The accident scene was inspected and the roadway geometry was mapped.
2. Physical evidence still present at the time of this inspection was also mapped.
3. The survey was used to create an accident scene diagram in a CAD program.
4. The striated tire marks were located on the diagram using photogrammetry. This process included locating the tire marks and, also, determining the striation directions.
5. A computer model of the vehicle was created based on vehicle specifications.
6. The vehicle model was positioned on the tire marks in the diagram.
7. The striation analysis was then conducted.

Figure 7.13 depicts the motion of the vehicle as defined by the tire mark evidence. The rest position of the vehicle is also shown. Figure 7.14 depicts a closer view of the vehicle as it passed across the yellow line. Figure 7.14 also includes the rectified photograph aligned with the accident scene diagram. Figure 7.15 focuses on the rear left tire at the time it deposited a striated tire mark on the yellow line.

FIGURE 7.13 Motion of the vehicle.

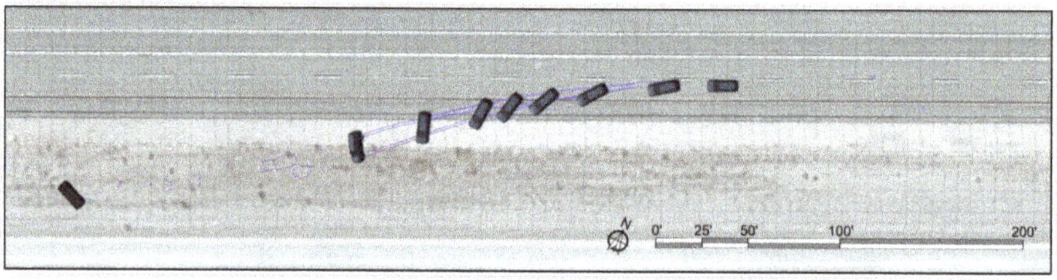

FIGURE 7.14 Position of vehicle as its left rear tire crossed the fog line.

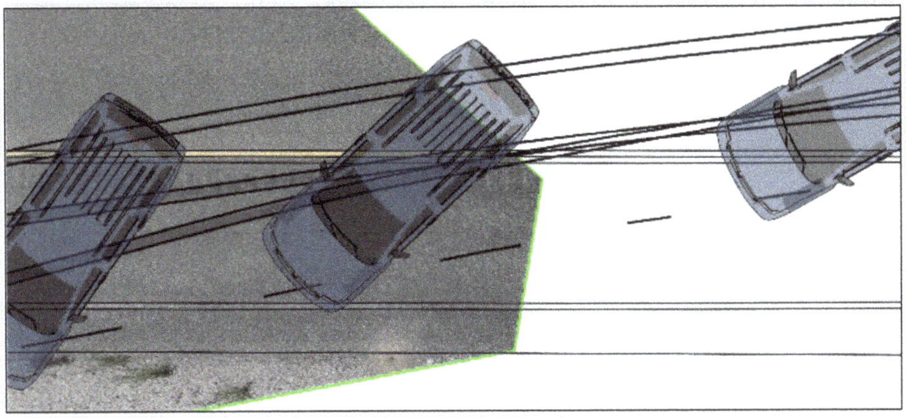

FIGURE 7.15 Left rear tire on the striated tire mark along with the measurements of α and θ.

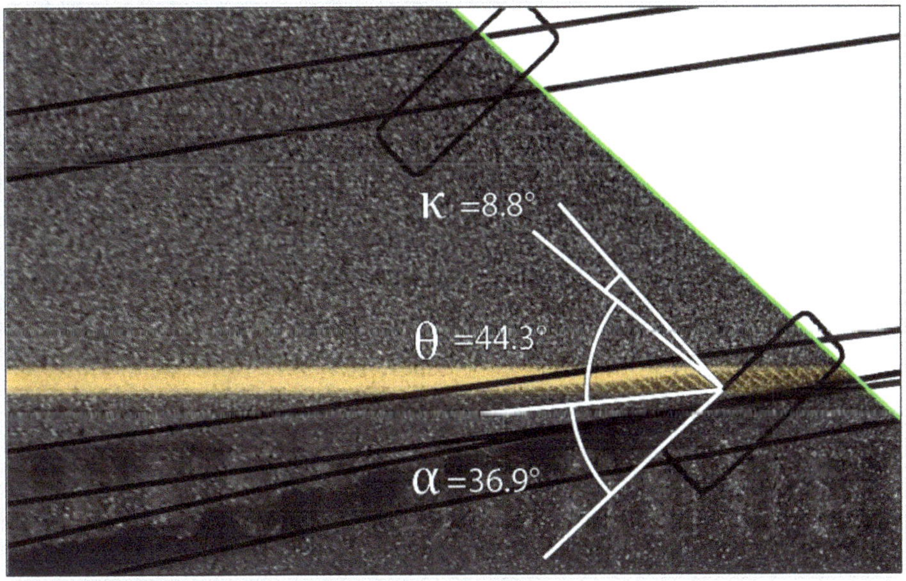

In the absence of braking, the striation marks would be aligned perpendicular to the tire heading. Therefore, steering the tires changes the orientation of the striation marks. When assessing braking levels, the analysis should begin with the rear tires and rear tire marks since the rear tires cannot be steered. The tire mark deposited by the rear left tire is the focus of this case study.

Once the vehicle model was aligned with the tire mark evidence, the angles α (36.9°) and θ (44.3°) were measured. The uncertainties of the tire heading, velocity direction, and striation direction were approximated. The uncertainties were estimated by observing what range of angles were visually consistent with the physical evidence. A ±2° range of tire headings still positioned the vehicle on the tire mark evidence, a ±1° range fit the velocity direction, and a ±0.5° range fit with the striation direction. In this case, the longitudinal slip was determined to be 12 ± 2%. This means that the driver was applying the brakes when these marks were deposited. In analyzing the speed of the vehicle, the braking can then be quantified in the form of longitudinal slip. The computed longitudinal slip values can then be used in the speed analysis.

References

1. Lambourn, R.F., "The Calculation of Motor Car Speeds from Curved Tyre Marks," *Journal of the Forensic Science Society* 29 (1989): 371-386, doi:10.1016/S0015-7368(89)73284-9.
2. Yamazaki, S., "Friction Coefficients of Tires and Skid Marks Left by Tires," *International Workshop on Traffic Accident Reconstruction*, 1998.
3. Gardner, J.D. and Moffatt, E.A., *Controlled Studies Illustrating Tiremarks Generated by Radial Tires*, (New York: The American Society of Mechanical Engineers, 1980), 35-53.
4. Yamazaki, S. and Akasaka, T., "Buckling Behavior in Contact Area of Radial Tire Structures and Skid Marks Left by Tires," *JSAE Review* 9, no. 3 (July 1988): 51-55, ISSN 0389-4304.
5. Baker, K.S., *Traffic Collisions Investigation*, (Evanston, IL, Northwestern University Center for Public Safety, 2001), ISBN: 0912642092.
6. Ishikawa, H., "Computer Simulation of Automobile Collision-Reconstruction of Accidents," SAE Technical Paper 851729, 1985, doi:10.4271/851729.
7. Fricke, L.B., *Traffic Crash Reconstruction*, 2nd ed., (Evanston, IL: Northwestern University Center for Public Safety, 2010), ISBN 0912642033.
8. Beauchamp, G., Hessel, D., Rose, N., Fenton, S. et al., "Determining Vehicle Steering and Braking from Yaw Mark Striations," *SAE Int. J. Passeng. Cars - Mech. Syst.* 2, no. 1 (2009): 291-307, doi:10.4271/2009-01-0092.
9. Beauchamp, G., Thornton, D., Bortles, W., and Rose, N., "Tire Mark Striations: Sensitivity and Uncertainty Analysis," *SAE Int. J. Trans. Safety* 4, no. 1 (2016a): 121-127, doi:10.4271/2016-01-1468.
10. Reveley, M., Brown, D., and Guenther, D., "A Comparison Study of Skid and Yaw Marks," SAE Technical Paper 890635, 1989, doi:10.4271/890635.
11. Fricke, L.B., *Traffic Accident Reconstruction - Volume 2 of the Traffic Accident Investigation Manual*, (Evanston: Northwestern University Center for Public Safety, 1990), ISBN 978-0912642079.
12. Hague, D.J., Lambourn, R.F., and Turner, D.F., *Critical Speed Studies I: The Accuracy of Speeds Calculated from Critical Curve Marks, and Their Striations*, (Telford, England, Institute of Traffic Accident Investigators, 1997).
13. Hague, D.J., Turner, D.F., and Williams, A.J., *Critical Speed Studies II: The Generation of Tyre Marks by Cornering Vehicles*, (Telford, England, Institute of Traffic Accident Investigators, 1997).
14. Beauchamp, G., Pentecost, D., Koch, D., and Rose, N., "The Relationship between Tire Mark Striations and Tire Forces," *SAE Int. J. Trans. Safety* 4, no. 1 (2016): 134-150, doi:10.4271/2016-01-1479.

CHAPTER 8

Error Rates of the Speed Analysis Methods

The full-scale rollover crash tests reported by Asay [1], Asay [2, 3], Stevens [4], and Larson [5] allow us to assess the accuracy of the rollover reconstruction techniques presented so far in this book. In each of these chapters, test data was used to develop the models, so a reasonable level of accuracy would be expected. However, it remains to examine these models in combination. How well do these models perform when combined to analyze an entire rollover crash? What is the expected error rate for speed calculation carried out with these models? This chapter addresses these questions.

Reconstructions of the Crash Tests

Seventeen of the tests were reconstructed. In fifteen of the tests, the vehicle rolled over off-road. In the two remaining tests, the vehicle began rolling over on the road and finished rolling off-road. For each Asay test, the survey data was used to generate a physical evidence diagram which included roadway slope. For Stevens' and Larson's tests, the diagrams from the original article were scaled and used for the analysis. Vehicle positions were aligned to the physical evidence in each diagram. In the Stevens' and Larson's tests, the ground was assumed to be flat. Accounting for these path slopes could change the results slightly, but not to a degree that would be expected to change the trends discussed below.

Rollover Phase – The focus of this analysis was calculating the translational speed at the beginning of the roll phase. Roll rates were not computed. An average deceleration rate of 0.44g was used for the roll phase of all the tests.

Trip Phase – For the tests in which the vehicle rolled off-road, the trip model presented in Chapter 5 was used. The beginning of the trip phase was defined as where the last trailing side tire lifted off, and ending at the termination of the tire marks, where the roll phase began. While other definitions of the trip phase were considered, this definition was found to produce the best overall accuracy. For the tests where the vehicle rolled on-road, the trip model was not used. Instead, the trip phase was modeled by extending the loss-of-control model to the end of the tire marks where the roll phase began. This was done because the trip model can result in lateral forces above the level that can be generated by the tires on the pavement. By using the loss-of-control phase model for the trip phase, the tire-roadway friction establishes the force limits. Comparatively, when a vehicle rolls off-road, the tires can furrow into the soil so high lateral forces are expected.

Loss-of-Control Phase – The loss-of-control phase was reconstructed using the CRASH model presented in Chapter 6. The sliding friction of the paved surface was set to 0.76 for each test. Using a friction value of 0.76 would be typical in the absence of any scene-specific friction testing data. For the off-road surface, the sliding friction was set to 0.6. For on-road rollovers, the CRASH model was used up until the beginning of the roll phase. Figures 8.1 to 8.17 depict scene diagrams of each test. The speed at the beginning of the physical evidence was reconstructed and compared to the initial speed of the vehicle.

Table 8.1 summarizes the results of the reconstructions. A comparison of the reconstructed and actual speeds is made at the beginning of each phase. The actual speeds at the beginning of rollover are the corrected speeds presented by Arndt [6]. For the several tests, the speeds at the beginning of the trip phase were not known, so that comparison was not made. As expected, the reconstruction of the roll phase using the average deceleration rate underestimated the actual speed in some tests and overestimated the actual speed in others. The reconstructed speed of the vehicle at the beginning of the evidence tended to underestimate the actual initial speed in all cases.

FIGURE 8.1 Asay [1] Test #2 – 1996 Buick Skylark.

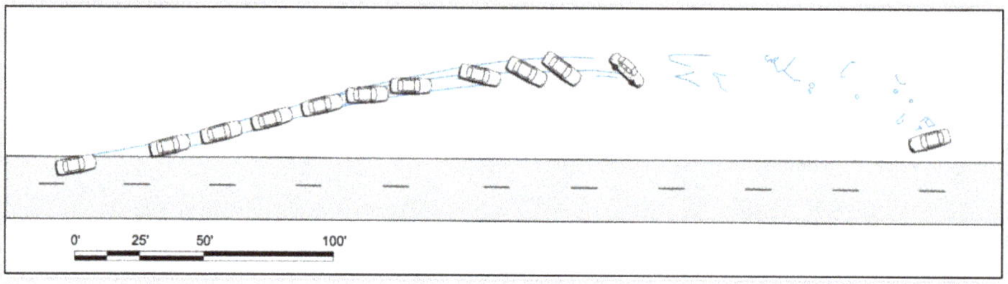

FIGURE 8.2 Asay [1] Test #3 – 1984 AMC Eagle.

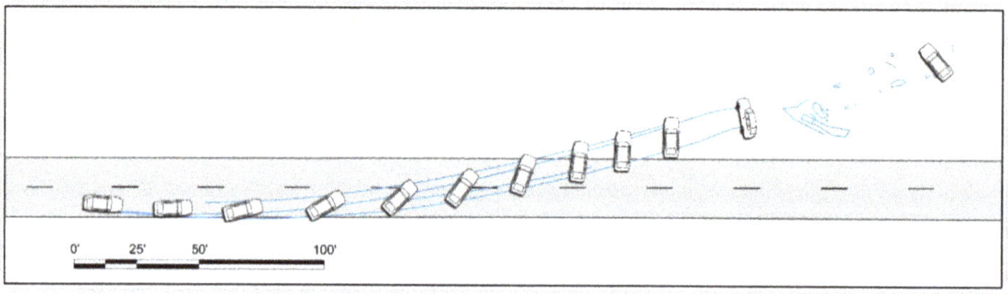

FIGURE 8.3 Asay [1] Test #5 – 1987 Ford Taurus.

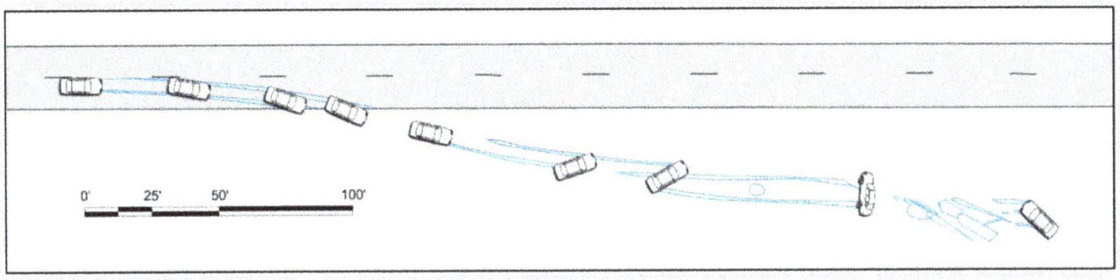

FIGURE 8.4 Asay [1] Test #6 – 1991 Ford Escort.

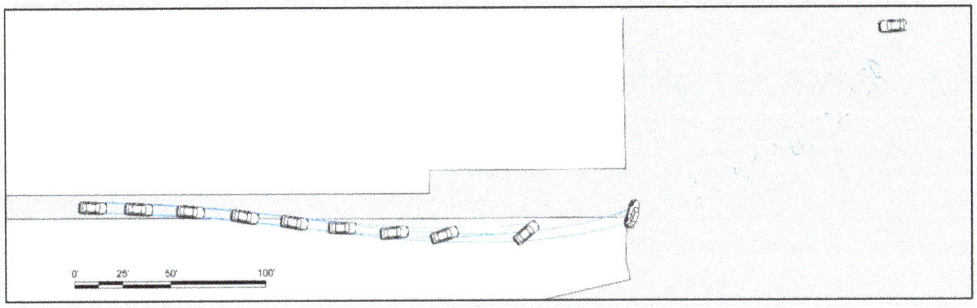

FIGURE 8.5 Asay [2] Test #1 – 1996 Oldsmobile Bravada.

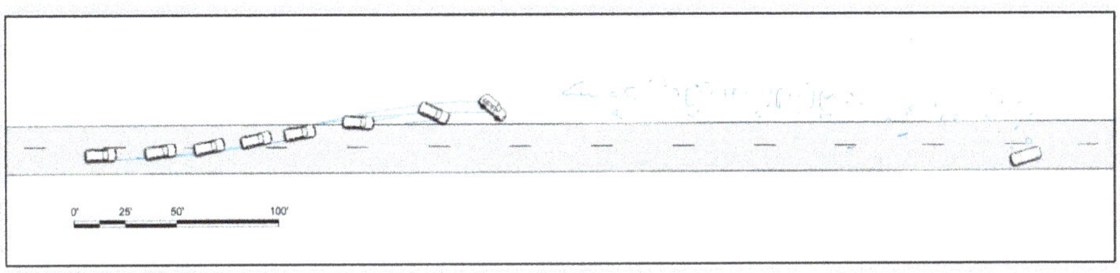

FIGURE 8.6 Asay [2] Test #2 – 1991 Isuzu Rodeo.

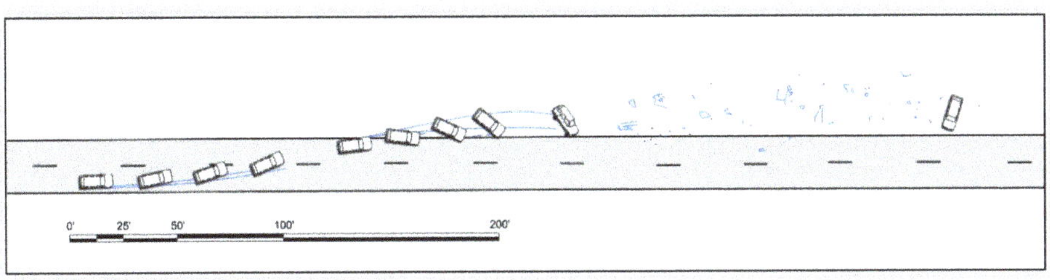

FIGURE 8.7 Asay [2] Test #4 – 1994 Nissan Pathfinder.

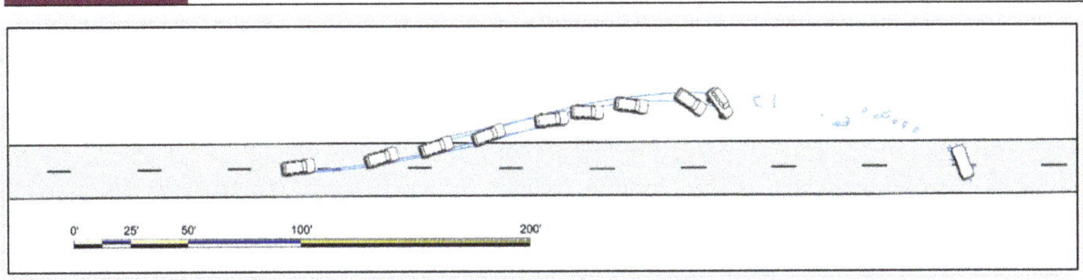

FIGURE 8.8 Asay [2] Test #6 – 2002 Ford Explorer.

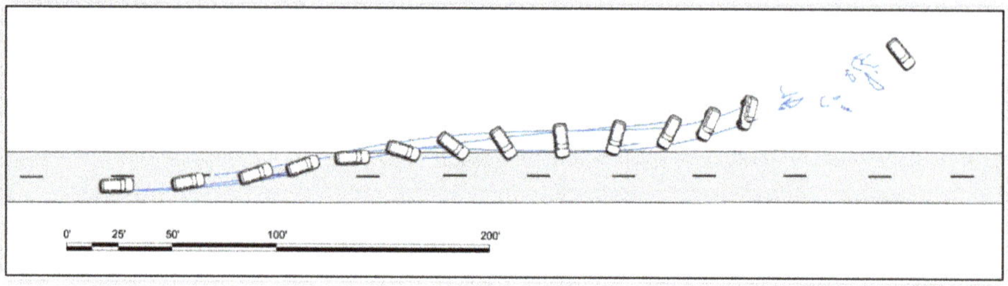

FIGURE 8.9 Asay [2] Test #7 – 1998 Ford Expedition.

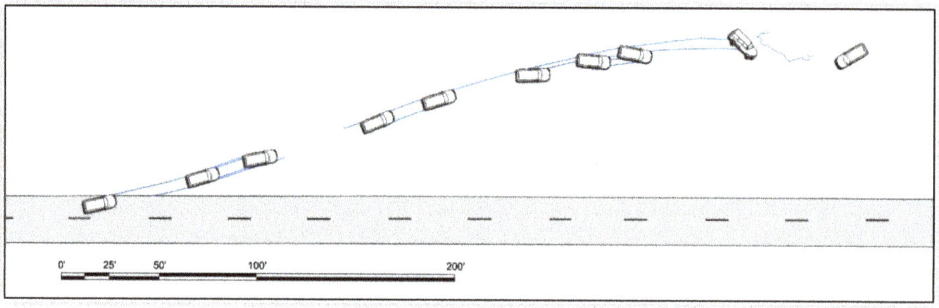

FIGURE 8.10 Asay [2] Test #8 – 1991 Mitsubishi Montero.

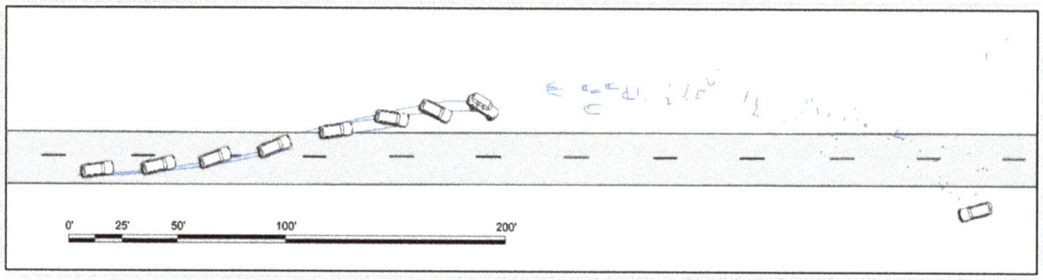

FIGURE 8.11 Stevens [4] Test #1 – 2001 Chevrolet Blazer.

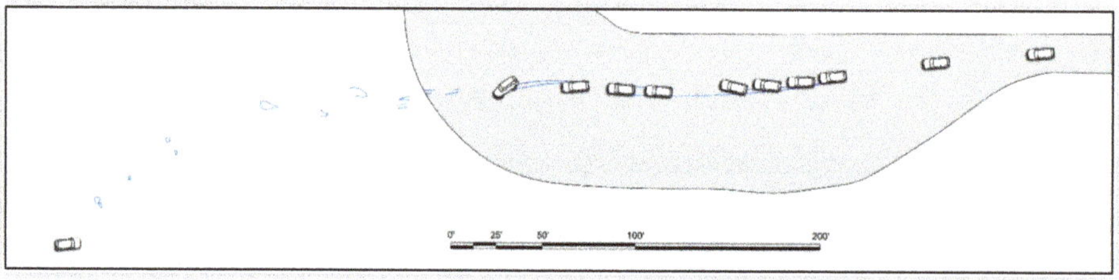

FIGURE 8.12 Stevens [4] Test #2 - 2002 Ford Explorer.

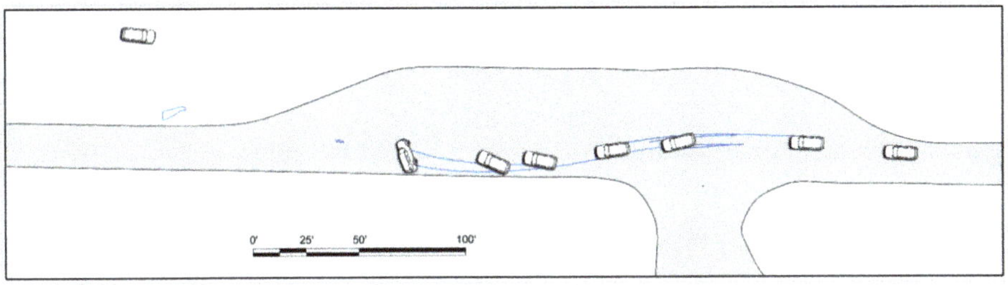

FIGURE 8.13 Stevens [4] Test #3 - 1997 Ford Explorer.

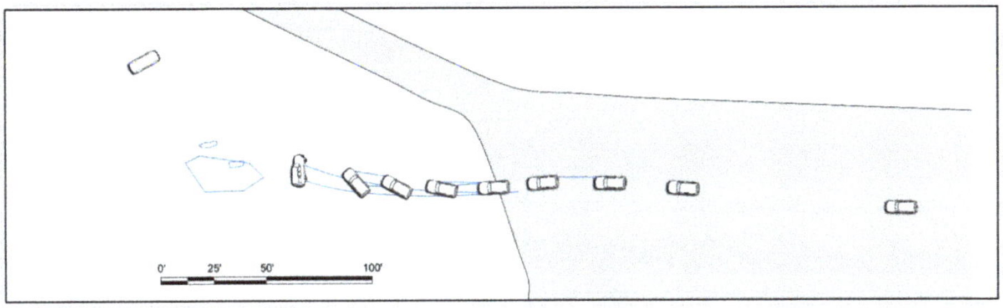

FIGURE 8.14 Stevens [4] Test #4 - 1995 Nissan Pathfinder.

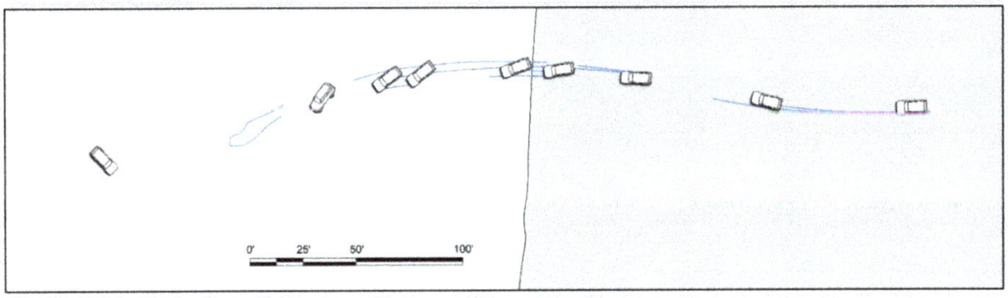

FIGURE 8.15 Asay [3] Test - 1996 GMC Jimmy.

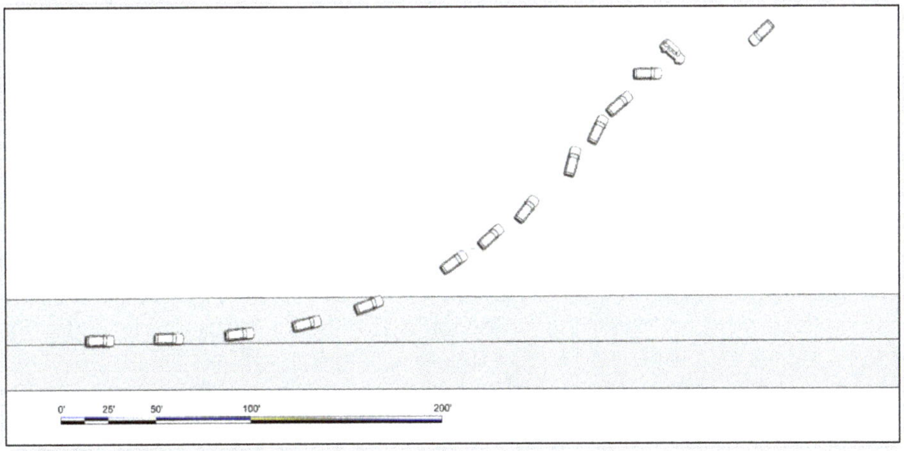

FIGURE 8.16 Larson (2015) - Toyota Tundra.

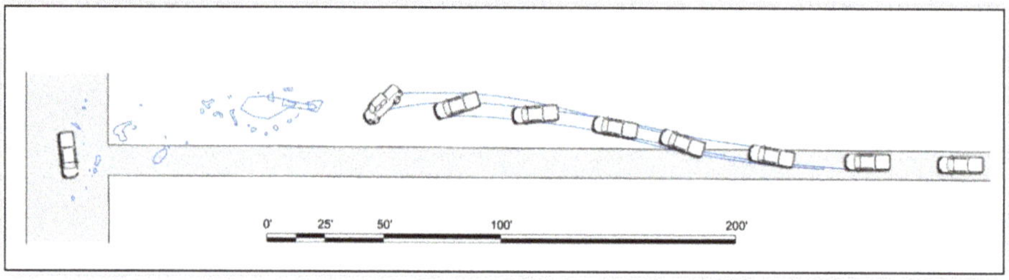

FIGURE 8.17 Larson (2015) - Toyota Camry.

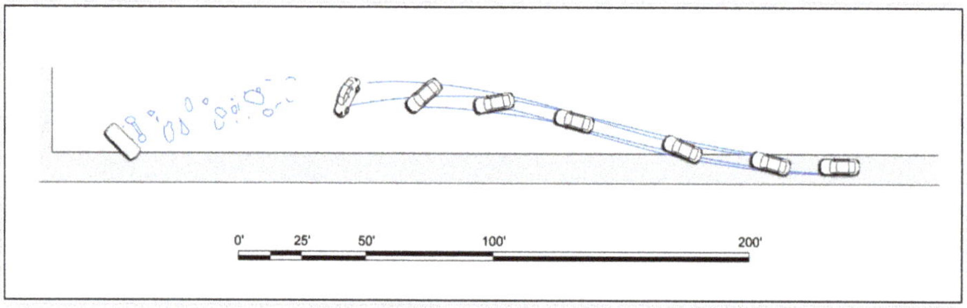

Discussion of Reconstruction Results

In Table 8.1, the speed at the beginning of physical evidence was compared to the initial speed of the vehicle. The comparison was made in this way because the initial speed of the vehicle is typically of interest, and the beginning of the physical evidence is typically the earliest speed that can be calculated. In real-world crashes, there can be speed lost before physical evidence due to driver steering or braking maneuvers, or other factors that slow the vehicle but do not deposit visible evidence on the roadway. In each of these tests, there wasn't any braking before the beginning of the evidence, but the vehicle did develop some slip angle before the tires start depositing evidence. There is some speed loss from this build up slip angle, although it is minimal, due to the short duration of the slip development phase and low slip angle that the tire marks appear in each test. While it may contribute in a minimal way, the underestimates in speed are not due primarily to the speed being calculated at the beginning of evidence. In some real-world cases, the speed at beginning of evidence may be significantly different than the initial speed. If the slip angle of the vehicle at the beginning of the physical evidence is large, for example.

The models used to analyze the trip and loss-of-control phases in this exercise did not account for any steering. The steering angles were large throughout the majority of these tests. Steering had the effect of increasing the slip angle of the front tires relative to the vehicle sideslip angle. In Asay's 2009 tests, the second steering input rapidly turned the tires to their limit. As a result of the steering not being incorporated, the models underestimated the speed at the beginning of the physical evidence, even when the speed at the beginning of the rollover phase was overestimated.

Asay's 2010 Test #6 was unique in the way the vehicle yawed and deserves additional attention. Figure 8.18 depicts the motion of the vehicle in this test. Unlike the other tests, the Explorer yawed beyond a 90° sideslip angle so that the rear of the vehicle was leading prior to rolling over. As the vehicle began to yaw, the slip angle of the front tires was more than the sideslip angle of the vehicle. However, after the vehicle yawed past 90° and was rear leading, the steering to the right resulted in the slip angles of the front tires being less than the sideslip angle of the vehicle. In this case, the reconstruction of the roll phase yielded

TABLE 8.1 Reconstruction results for the Asay and Stevens full-scale rollover tests

Reconstruction vs test data														
			Roll phase				Trip phase				Yaw phase			
			Actual rollover speed	Calculated rollover speed	Speed at beginning rollover difference (mph)		Actual speed at beginning of trip	Calculated speed at beginning of trip	Speed at beginning of trip difference (mph)		Actual initial speed	Calculated speed at beginning of evidence	Speed at beginning of evidence difference (mph)	
Test no.	Trip	Vehicle	(mph)	(mph)	(mph)	(%)	(mph)	(mph)	(mph)	(%)	(mph)	(mph)	(mph)	(%)
Asay 2	Off Road	1996 Buick Skylark	41.6	39.2	-2.4	-6%					75	53.5	-21.5	-29%
Asay 3	Off Road	1984 AMC Eagle	34.6	28.7	-5.9	-17%					75	70.1	-4.9	-7%
Asay 5	Off Road	1987 Ford Taums	27.1	29.1	2	7%					75	68.9	-6.1	-8%
Asay 6	Off Road	1991 Ford Escort	51.6	47.7	-3.9	-8%					75.7	69.5	-6.2	-8%
Asay 1	Off Road	1996 Oldsmobile Bravada	57.2	59.2	2.0	3%					75.7	68.5	-7.2	-10%
Asay 2	Off Road	1991 Isuzu Rodeo	44.4	47.3	2.9	7%	60.7	58.1	-2.6	-4%	73.5	64.5	-9	-12%
Asay 4	Off Road	1994 Nissan Pathfinder	36.7	38.3	1.6	4%	52.3	47.2	-5.1	-10%	69.90	50	-19.9	-28%
Asay 6	Off Road	2002 Ford Explorer	31.1	30.8	-0.3	-1%	43.2	47.7	4.5	10%	70	68.6	-1.4	-2%
Asay 7	Off Road	1998 Ford Expedition	28.7	26.7	-2.0	-7%	52.1	41.2	-10.9	-21%	70	44.8	-25.2	-36%
Asay 8	Off Road	1991 Mitsubishi Montero	61	55.4	-5.6	-9%	70.3	61.3	-9.0	-13%	74.6	63.8	-10.8	-14%
Stevens 1	On Road	2001 Chevrolet Blazer	57.9	58.9	1	2%					71	66	-5	-7%
Stevens 2	On Road	2002 Ford Explorer	41	44	3	7%					58	54.3	-3.7	-6%
Stevens 3	Off Road	1997 Ford Explorer	30.2	34.7	4.5	15%					55.5	51.3	-4.2	-8%
Stevens 4	Off Road	1995 Nissan Pathfinder	42.3	37.4	-4.9	-12%					62	55.7	-6.3	-10%
Asay	Off Road	1996 GMC Jimmy	20.1	25.9	5.8	29%	31.4	37	5.6	18%	53.2	44.3	-8.9	-17%
Larson Pickup	Off Road	Toyota Tundra	43.6	42.3	-1.3	-3%					63.4	55.4	-8	-13%
Larson Sedan	Off Road	Toyota Camry	33.7	34.3	0.6	2%					61.9	57.4	-4.5	-7%

FIGURE 8.18 Rollover motion from Asay Test #6.

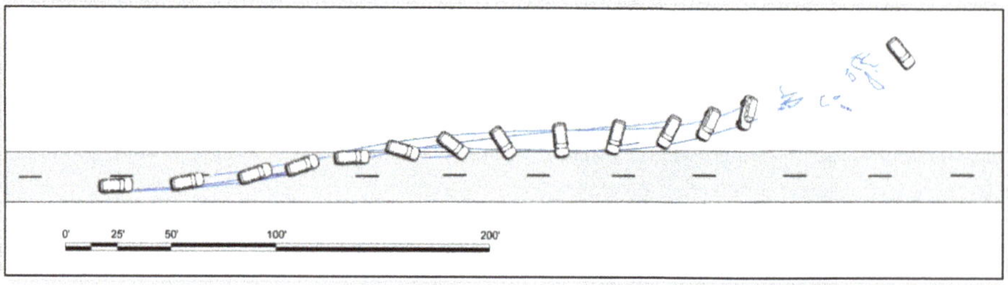

31 mph, the same as the actual speed. The trip phase reconstruction overestimated the speed during that phase, consistent with the rear-leading yaw and right steering input. The reconstruction then underestimated the speed loss during the loss-of-control phase, again consistent with the steering input. The overestimation during the trip phase and underestimation during the loss-of-control phase had a canceling effect. In fact, the reconstructed speed at the beginning of the evidence was closest to the actual speed for Asay Test #6.

The models discussed in this chapter tended to underestimate the speed of the vehicle at the beginning of the physical evidence on the roadway. The models for the trip phase and loss-of-control phase used in this chapter considered the sideslip angle of the vehicle, not the slip angle of the individual tires. In the tests, the final steer was held until the vehicle rolled over. In many of these tests, the final steer was the maximum possible steering angle. These large sustained steering inputs create a significant difference between front tire slip angle and vehicle sideslip angle, which contributed to the reconstructions underestimating the speed at the beginning of the evidence.

The error rates presented in Table 8.1 are dependent, in part, on the specific steering inputs in those tests. In many real-world rollover crashes, the driver does not hold the steering wheel all the way to one side until the vehicle rolls over, as occurred in many of the crash tests presented here. Smaller steering inputs can also result in vehicle rollover. Smaller steering inputs would reduce the difference between the slip angle of the front tires and the sideslip angle of the vehicle and would therefore, all things being equal, reduce the amount of error in the calculations. In many cases drivers would have good reason to countersteer prior to the vehicle rolling over. The tire mark striations may show that a countersteer did occur. Countersteering would reduce the slip angles of the front tires and therefore reduce the amount of error in the calculations.

While the steering could be accounted for within the models for the trip and loss-of-control phases – different slip angles could be used for the front and rear tires – access to the front steering angles is typically not available or limited to isolated instances where striations are visible in the deposited tire marks.

The actual motion of the vehicle is likely a better indicator of the tire steer angles, which is not determined by the spreadsheets used in this exercise. An approach that could potentially be used to incorporate the steering inputs is simulation, which is discussed in the next chapter.

References

1. Asay, A. and Woolley, R., "Rollover Testing on an Actual Highway," *SAE Int. J. Passeng. Cars - Mech. Syst.* 2, no. 1 (2009): 1237–1267, doi:10.4271/2009-01-1544.
2. Asay, A. and Woolley, R., "Rollover Testing of Sport Utility Vehicles (SUVs) on an Actual Highway," SAE Technical Paper 2010-01-0521, 2010, doi:10.4271/2010-01-0521.

3. Asay, A., Carter, J., Funk, J., and Stephens, G., "Rollover Testing of a Sport Utility Vehicle (SUV) with an Inertial Measurement Unit (IMU)," SAE Technical Paper 2015-01-1475, 2015, doi:10.4271/2015-01-1475.
4. Stevens, D., Arndt, S., Wayne, L., Arndt, M. et al., "Rollover Crash Test Results: Steer-Induced Rollovers," SAE Technical Paper 2011-01-1114, 2011, doi:10.4271/2011-01-1114.
5. Larson, R., Croteau, J., Bare, C., Zolock, J. et al., "Steering Maneuver with Furrow-Tripped Rollovers of a Pickup and Passenger Car," SAE Technical Paper 2015-01-1477, 2015, doi:10.4271/2015-01-1477.
6. Arndt, M., "Drag Factors from Rollover Crash Testing for Crash Reconstructions," Proceedings of the ASME 2011 International Mechanical Engineering Congress & Exposition IMECE2011, Denver, CO, November 11-17, 2011, IMECE2011-65537.

CHAPTER 9

Advanced Analysis Methods: Simulation

Introduction

A number of the models discussed in the previous chapters can be extended to a three-dimensional form with the use of simulation software, such as PC-Crash or HVE. This chapter explores the use of PC-Crash and HVE for modeling each of the phases of a single-vehicle rollover crash. A balanced and realistic approach toward simulation is adopted here - neither embracing nor rejecting simulation out of hand. PC-Crash and HVE both have potential weaknesses when it comes to modeling rollover crashes, but a blanket dismissal of these software packages would be ill considered. As the statistician George E.P. Box has said, "all models are wrong, some are useful."[10] Both PC-Crash and HVE are certainly useful! In these authors experience, many reconstructionists dismiss simulation when their early efforts with a particular software package do not yield fruit. Some analysts blame their own lack of skill on the software. In giving up too easily, many reconstructionists miss out on the benefits of simulation for integrating and extending the models discussed in previous chapters. This chapter demonstrates how these models can be integrated and extended with the use of PC-Crash and HVE.

Both PC-Crash and HVE are physics based, but their validity for any particular application relates to the quality of the models they use to calculate the magnitude and temporal variation of the forces applied to the vehicle. Users of these software packages should understand how the models within each software package work. As the PC-Crash website has stated: "PC-Crash is technical software for serious users. It's powerful and gives you a lot of freedom, so that you can simulate unique events. It therefore necessarily leaves you the option to make a mess. Don't let yourself do this." This statement is also applicable to HVE.

[10] Wikipedia, *George E. P. Box*, http://en.wikipedia.org/ wiki/George_E._P._Box (as of Dec. 9, 2008, 20:12 GMT).

The sensitivity of a simulation to various model inputs within both PC-Crash and HVE clearly depends on the collision type, and the discussion here is confined to rollover modeling. As an example, the suspension model within these software packages are clearly more influential for rollover modeling than for modeling planar collisions. Even within rollover modeling, the suspension model matters more when the user is attempting to predict a rollover than when they are reconstructing a rollover that has already occurred. PC-Crash or HVE could be useful for reconstructing the deceleration-time history for a rolling vehicle within the context of accident reconstruction. That would not necessarily imply that they are adequate – or that they need to be adequate – for use as a fully predictive rollover dynamics model.

Before delving deeper into the use of simulation to model rollover crashes, it is helpful to define several terms related to simulation validation. The purpose in defining these terms is simply to give the reader a framework for interpreting prior studies related to the simulation of rollover crashes and a framework for knowing the limitations of simulation for simulating rollovers. Our purpose in this chapter is not a full validation of either PC-Crash or HVE. This chapter does, though, give the reader a sense for the degree to which PC-Crash and HVE have been validated for various tasks involved in rollover reconstruction. The definitions and concepts presented here are drawn from articles by Kleijnen [1], Balci [2], Robinson [3], Carson [4], and Sargent [5].

Conceptual Model Validation: All simulation programs and reconstruction models use *conceptual models* that attempt to mimic real-world systems. Examples would be the tire and suspension models in PC-Crash and HVE. The validity of these conceptual models could be evaluated and demonstrated independent of their implementation in PC-Crash, a process that can be termed *conceptual model validation*. An example would be to compare the PC-Crash tire model to real-world tire data to determine the degree to which the model accurately mimics the behavior of real tires. One important idea to keep in mind here is that the degree to which the model needs to mimic the real-world system in order to be valid depends on the application to which it is being applied.

Calibration: Conceptual model validation could also relate to model *calibration* where the validity of the model is assumed to have been demonstrated, and its use is a matter of using physically realistic inputs to obtain correct results. An example of this would be to use tire data from the literature to determine reasonable inputs into the PC-Crash TM-Easy tire model.

Verification: Each of the conceptual models used by PC-Crash and HVE also has its implementation within the software, through the computer code underlying the software. The process of ensuring that these conceptual models are correctly implemented in the code is referred to as *verification*. Verification aims at eliminating any programming errors. Most validation studies of PC-Crash and HVE assume the software developers have adequately addressed this task.

Operational Validation: Operational validation involves ensuring that the implemented conceptual models, when they are all combined in the software, produce a sufficient level of accuracy when compared to the real-world process or system the model attempts to replicate. Carson [4] noted that "sufficient accuracy means that the model can be used as a substitute for the real system for the purposes of experimentation and analysis…" Robinson [3] notes that "a key concept is the idea of sufficient accuracy. No model is ever 100% accurate…the aim is to ensure that the model is sufficiently accurate… this accuracy is with reference to the purpose for which the model is to be used." Again, here, sufficient accuracy would be defined relative to the purpose for which the model is utilized. Not only that, sufficient accuracy can also be defined in relationship to other available models.

In relationship to operational validation, another issue that should be acknowledged is the degree to which user skill can affect the simulation results. Some analysts are more skilled than others. This is true not only of analysis conducted within PC-Crash but of any simulation software or reconstruction technique. This is an important issue for validation because not all studies that explore the validity of PC-Crash have isolated the validity of the physical models in PC-Crash from the skill of the user applying those models. In other words, results of some validation studies may tell us as much about the skill level of the user than they do about the validity of PC-Crash. These studies are still useful, but it should be recognized that the underlying models might be stronger and more robust than what any particular user is able to demonstrate.

Day [6] raised this same issue, noting that "misuse [of computer programs for accident reconstruction] is due to the lack of a thorough understanding of how the programs work...Just as the level of skill varies among investigators, the level of understanding how the programs work also varies. When properly used, these computer programs are an invaluable accident investigation tool. When misused, these programs can produce erroneous results – and a misconception of what actually occurred during the accident."

Wach [7] also raises this issue, noting that "it should be emphasized that PC-Crash is only a tool which gives correct results in so far as the data are correct and mathematical models for the investigated physical phenomena are relevant, therefore the final responsibility for the conclusions derived from calculations rests with the user...to use the program correctly, knowledge of the dynamics of vehicles and collisions is indispensable..."

An example of demonstrating the operational validity of PC-Crash would be to compare simulations of crash tests to the measured data from the crash test. One issue that arises when assessing the operational validity of PC-Crash is the validity of the data input into the software. One way to pose the question of operational validity would be, "If we know that all of the inputs are valid, then how well do the models within PC-Crash mimic the dynamics of a real-world crash?" This question can be problematic because some parameters, such as the coefficient of restitution and the intervehicular friction, are not directly measurable from a crash test and calculation of these parameters can be difficult. Some validation studies also ask the evaluator to use values for input parameters that they would typically use in their accident reconstruction practice, rather than providing the actual known input values. This is one way of setting up a validation study and unavoidable to some extent. However, if a study is set up this way, it should be recognized that user skill is entering into the results to some degree. Another way to set up the study would be to provide the analyst with all the known data from the crash test in order to decrease (though certainly not eliminate) the degree to which the user's knowledge and skill enter into the results.

Credibility: Another issue related to validation is a simulation software's *credibility*. In a validation context, the term credibility refers to the confidence of the community of users in the simulation software. Wide use of a simulation software package like PC-Crash or HVE can be an indication that the relevant community accepts it and judges it valid. An indication of this would be the presence of studies in the literature that assume the validity of the simulation software.

Human Driver Steering Capability

During a typical high-speed, single-vehicle rollover crash, the loss of control is initiated by driver steering inputs. Thus, in simulating this type of rollover, the analyst typically enters steering inputs into the simulation to generate the loss of control. In a reconstruction context, the analyst may be attempting to match a set of tire marks, and the steering inputs (magnitude and rate) are varied until the evidence is matched. In determining reasonable inputs for a simulation, the analyst can consider the following data related to the rates at which human drivers can actually steer.

In a 1976 study, Rice [8] conducted testing with 90 nonprofessional drivers. These drivers were instructed to drive as aggressively as they could through a driving course while successfully traversing the course. The driving course, which included elements such as an off-road recovery, large radius arcs, an avoidance maneuver, and a surprise intrusion, was set up to require steering inputs greater than what would be required for typical driving. Rice reported steering rates for drivers traversing the avoidance maneuver portion of the course. He indicated that, for successful runs through this section, the average maximum steering rate was 520°/s. For unsuccessful runs through this section, the average maximum steering rate was 850°/s. Rice also studied 11 runs involving loss of control in other parts of the course and reported that these runs exhibited steering rates in excess of 1000°/s.

Mazzae [9, 10] examined driver crash avoidance behavior in a right-side intersection incursion scenario using field testing at a test track with actual vehicles and using the Iowa Driving Simulator. The field testing was conducted in the Vehicle Dynamics Area at the Transportation Research Center in East Liberty, Ohio. The study included 245 test subjects between the ages of 25 and 55 years, with 192 participating in dry pavement testing and 53 participating in wet pavement testing. Participants drove either a 1995 Chevrolet Lumina or a 1996 Ford Taurus. Unexpectedly, during each participant's drive, an artificial full-scale vehicle was pushed 6 ft into the path of their vehicle from the right. In the dry pavement study, 94% of the subjects attempted both steering and braking inputs, and in the wet pavement study, 98% of the subject attempted both steering and braking inputs. In both the dry and wet pavement studies, roughly half steered first and half braked first. The average maximum steering rate observed during the avoidance maneuvers was 262°/s for the dry pavement and 294°/s for the wet pavement. The highest observed steering rates were 1159°/s for the dry pavement and 1335°/s for the wet pavement study. Mazzae noted that "ninety-five percent of the steering rates observed were less than 600 degrees per second in the dry pavement study and less than 643 degrees per second in the wet pavement study."

The simulator study utilized 60 males and 60 females between the ages of 25 and 55 years. All 120 subjects steered and braked in an attempt to avoid a collision with a vehicle coming into their lane. Seventy-nine percent of the subjects applied the brakes first, prior to steering. Seventeen percent steered first. The remaining 4% steered and braked simultaneously. The average magnitude of the avoidance steering input was around 148° and the average maximum steering rate was 514°/s. Ninety-five percent of the steering rates were less than 981°/s. Around 35% of the subjects crashed in their attempt to avoid. Mazzae noted that, "steering inputs exhibited by subjects in this IDS study were larger and quicker than those observed in the related test track studies...This difference is believed to be attributable to the lack of 'road feel' present on the IDA as well as the limited range of travel of the simulator motion base." Nonetheless, Mazzae's results further establish the speed with which normal human drivers can turn a steering wheel.

Carr [11] conducted additional analysis of the data sets produced by Mazzae. For the dry pavement data, with ABS-equipped vehicles, Carr reported a 95th percentile peak steering input of 136° and a 95th percentile peak steering rate of 661°/s. Carr observed that the "driver choice of steering appears to be influenced by the amount of lateral acceleration experienced...The drivers seem to limit the amount of steering demand when the vehicle responds with moderate lateral acceleration or movement. When the vehicle does not have significant response, as in the situation of locked wheel braking [without ABS], the drivers seem to continue to apply more steering to try to achieve a desired lateral response." The typical three-phase high-speed rollover crash is likely the exception to this statement. Carr observed that "all of the yaw rates created as a result of the drivers' demands are well below the capacity of modern vehicles which generally exceed 35 to 45 degrees per second with no challenges to lateral stability on dry pavement surfaces." In other words, none of the drivers in the Mazzae study experienced the type of loss of control that typically precedes a single-vehicle rollover like the one depicted in Figure 1.1. This type of rollover crash often involves a driver steering their vehicle to the point that the vehicle capabilities are fully utilized and a complete loss of control occurs.

Bartlett [12] reported a study of driver capabilities documented in a dry asphalt, closed, cone-marked course. This study used 467 drivers (418 males and 49 females) who drove one of two police vehicles – a front-wheel drive 1996 Chevrolet Lumina and a rear-wheel drive 1992 Chevrolet Caprice. The drivers were instructed to drive "as if they were responding to assist another officer in trouble, with approval from their supervisors to drive as quickly as they could to facilitate his rescue." The participants drove through a series of maneuvers including a 100 ft radius circle, a decreasing radius turn, and a slalom course. Bartlett reported an average maximum steering rate of 828°/s (calculated over 100 ms). The maximum steering rate observed over 100 ms was 1199°/s for the male drivers and 919 for the female drivers.

In 2005, Forkenbrock [13] presented data related to the steering rates of drivers during double lane change tests conducted as a part of the National Highway Traffic Safety Administration's (NHTSA) Light Vehicle Handling and Electronic Stability Control (ESC) Effectiveness Research Program. These tests utilized five vehicles, all of which were equipped with electronic stability control (ESC). Testing was conducted with and without the ESC enabled. In examining this data, Forkenbrock documented instantaneous steering rates as high as 1819°/s. Even when averaged over a 1 s time period, Forkenbrock found that one driver sustained a steering wheel rate of 963°/s. Figure 9.1 shows the peak steering rate observed by Forkenbrock for each vehicle when the data was averaged over a 500 ms time period.

FIGURE 9.1 Steering rate data from Forkenbrock [13].

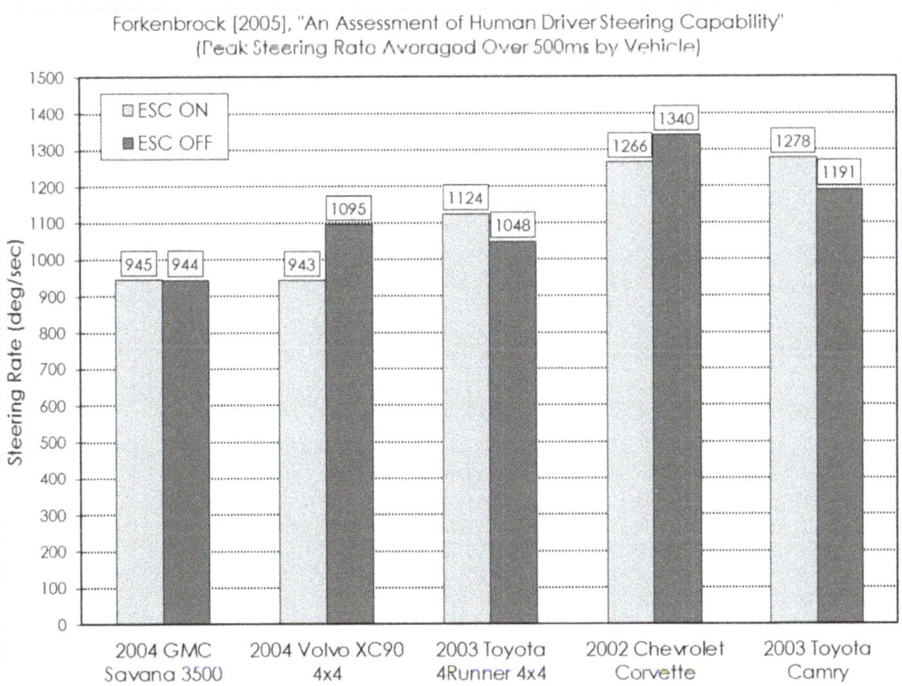

Forkenbrock noted that when NHTSA conducted the double lane change tests that he analyzed, it had not been anticipated that the data would be used to examine the steering capabilities of human drivers. He stated that the "emphasis was on path following…For this reason, the results…should not be taken to represent the absolute limit of a human performance. Rather, the authors present this information to help guide those developing maneuvers performed with automated steering; the data contained in this paper can be used to help researchers determine whether the steering demands imposed by a prospective test maneuver are reasonable."

The magnitude of the steering inputs and rates observed by Forkenbrock could also have been affected by the physical characteristics and experience level of the drivers. Forkenbrock reported that the double lane change maneuvers were performed by four experienced drivers, all four of which were male and of average build. Two of the drivers were in their twenties, one was in his thirties, and the fourth was in his sixties. Each driver had past experience as a test driver. Given the human capabilities data reviewed here, the steering rates used in the tests described above are severe, but physically possible.

Prior Studies Related to Simulating Rollover Crashes

HVE

Day and Garvey [14] evaluated the use of the EDVSM vehicle simulation model (a module within HVE) for analyzing rollover crashes. They discussed the vehicle model assumptions and presented several examples of the use of EDVSM, including simulating the effect of steer phasing on vehicle rollover, simulating a dolly rollover test, and simulating a real-world crash involving an off-road rollover. Day had presented a prior validation of EDVSM in 1997. This earlier validation had included limit maneuvers and rollovers. However, the rollover simulations were terminated after the vehicles had rolled approximately 35-50° because the vehicles in the tests to which the simulations were being compared were equipped with outriggers. In addition to that, at the time of the 1997 validation, EDVSM did not include the capability to model vehicle body-to-ground contacts. The version of EDVSM discussed by Day and Garvey in 2000 incorporated the capability of modeling these vehicle body-to-ground contacts.

For the dolly rollover simulation, Day and Garvey used EDVSM to simulate a dolly rollover of a 1981 Dodge Challenger. In this test, the vehicle was situated on the dolly at an initial roll angle of 28°. The dolly was accelerated up to a speed of approximately 30 mph before being suddenly decelerated. The vehicle exited the cart and began rolling as a result of its interaction with the dolly, and as a result of its interaction with the ground upon exiting the dolly. Day and Garvey reported that the results of their simulation for this test "compare quite favorably…For example, the total distance traveled for the test was 71 feet, compared with 80 feet in the simulation; total number of rolls was 3 in the test, compared to 2-1/2 for the simulation (the test vehicle comes to rest on its tires, while the simulated vehicle came to rest on its roof)…the initial total velocity in the simulation was slightly higher than in the test. Perhaps this is partially responsible for the additional distance traveled. However, the simulated vehicle rolled *less* than the test vehicle, thus, there must be some other factor at work; perhaps higher suspension shock absorber rates in the simulation model. Experience has shown that vehicles with like-new shocks, such as those on vehicles in the EDC Vehicle Database, do not roll as many times – nor as violently – as vehicles with worn shocks. This behavior appears to be the result of damping in both the vertical and roll degrees of freedom; both of these modes are very active during multiple rollover events."

In discussing the results of their study, Day and Garvey stated: "Matching the detailed results from a simulation with real-world crash site evidence and vehicle damage is a time-consuming process. This is especially true if the goal is to match the vehicle damage patterns. As with any reconstruction involving vehicle rollover, significant skill is required in the proper interpretation of the evidence...Once the evidence is properly interpreted numerous simulations must be performed, adjusting key parameters until a satisfactory match is achieved. The most important parameter is the velocity at the start of the rollover. However, matching the vehicle damage patterns also requires careful selection of vehicle stiffness on various portions of the vehicle. The problem can become even more challenging for crashes involving multiple rolls that caused superimposed damage on the same portion of the vehicle."

PC-Crash

STUDIES DESCRIBING THE CONCEPTUAL MODELS

In 1996, Steffan and Moser [15] described the collision and trajectory conceptual models used by PC-Crash. The *trajectory* model, which would certainly come into play when simulating rollover crashes, utilizes the tire and suspension force models and can also account for trailer coupling forces. PC-Crash currently offers two tire models – the linear tire model and the TM-Easy tire model, though at the time of Steffan and Moser's publication, only the linear tire model was available.

Steffan [16] discussed the conceptual models within PC-Crash that are relevant to modeling rollover crashes (tire and suspension models, ground surface modeling, and the vehicle body-to-ground contact model) and then presented a basic validation of PC-Crash for modeling rollovers. This validation consisted of using PC-Crash to model two rollover crash tests, and then visually comparing the overall vehicle motion between the tests and the simulations. While Steffan obtained favorable visual agreement between the overall vehicle motion in the tests and the simulations, no comparisons were made between the translational and angular velocities and accelerations experienced by the test vehicles and those experienced by the vehicles in the simulations.

In a book published in 2012, Wach [7] described the conceptual models and assumptions used by PC-Crash, including the coordinate systems, the tire models, the antilock brake model, the electronic stability program (ESP) model, the trailer coupling force model, the impact model, the stiffness-based impact model, the mesh-based impact model, the soft-soil model, and the multi-body model. Wach also reports historical information about PC-Crash, noting that "PC-Crash is a program for road accident simulation. It's first version was created at the Institute of Mechanics of the University of Graz, Austria, in the early 1990s. The author of the original idea and basic physical model is Professor Hermann Steffan, in close cooperation with Wolfgang Neubauer and Dr. Andreas Moser. The program development is supported by Dr. Steffan Datentechnik in Linz. The program, a unique world-wide standard in road accident reconstruction, is available in twenty-two language versions."

OPERATIONAL VALIDATION/CALIBRATION STUDIES

In 2009, Rose [17] presented PC-Crash modeling of a dolly rollover test he reported in 2008 [18]. This modeling utilized PC-Crash 7.3. While more recent versions of PC-Crash have incorporated changes relevant to rollover modeling, certain methodological issues addressed by Rose have not changed. Specifically, he found that the dominant variables affecting the match between the PC-Crash simulation and the test were the vehicle-to-ground friction coefficients, the vehicle suspension stiffness and damping, and the car body restitution. Perhaps more significantly, Rose found that obtaining a match between the simulated and actual roll velocity histories led naturally to good agreement with the simulated and actual translational velocity histories. This is consistent with the discussion in Chapter 4, where it was demonstrated that a relationship exists between the roll velocity history experienced by a vehicle and the vehicle's deceleration rate history. This relationship is mediated through the vehicle-to-ground friction coefficient. This finding indicates that PC-Crash could potentially be used to reconstruct the deceleration history for a vehicle during a rollover.

To do this, the reconstructionist would first reconstruct the vehicle's roll motion spatially based on physical evidence. Then, the vehicle's speed at the beginning of the rollover could be calculated using a constant deceleration rate. The vehicle's initial roll rate could be estimated based on models available in the literature or based on simulation. Starting with these initial conditions, the motion of the vehicle could be simulated to match the spatial reconstruction. In generating a simulation that matched the reconstructed roll motion, one would inherently reconstruct the deceleration rate history [19]. The accuracy of such an approach would clearly depend on the accuracy of the underlying spatial reconstruction of the gross vehicle motion.

Rose used a physical evidence diagram as a background image when optimizing the PC-Crash simulation of the dolly rollover he analyzed. The simulated vehicle motion was judged visually based on how well it agreed with the physical evidence on this diagram. The simulated motion was also compared to the translational and rotational velocity time histories determined from video analysis. Figure 9.2 presents a side-by-side comparison of images from the crash test video and from the simulation. These images demonstrate that there was reasonable agreement between the gross vehicle motion in the crash test and the simulation. During the leading side roof-to-ground impact, though, the roll angle of the vehicle in the simulation lags behind the roll angle of the vehicle in the crash test. Later, during the trailing side roof impact, the roll angle of the vehicle in the simulation gets out ahead of the roll angle of the vehicle in the crash test. The vehicle in the simulation also appears to penetrate into the ground more than what is warranted by the actual vehicle deformation during the crash test.

Andrews [20] published a study at the Enhanced Safety of Vehicles Conference that compared modeling of a staged rollover collision using PC-Crash Version 7.3 to the actual measured and videoed vehicle motion. Andrews concluded that "PC-Crash was successfully utilized to reconstruct the staged rollover collision. The PC-Crash reconstruction showed the speed of the vehicle at the point of roll to be within 2.1 mph of the actual data. The number of rolls and the vehicle path during the yaw phase and the rollover phase were consistent between PC-Crash and the collected data."

Zebala [21] examined the antilock braking system (ABS) model used in PC-Crash Version 8.2 and compared it to the results of extreme braking tests run with ABS-equipped

FIGURE 9.2 Comparison of simulation with test video.

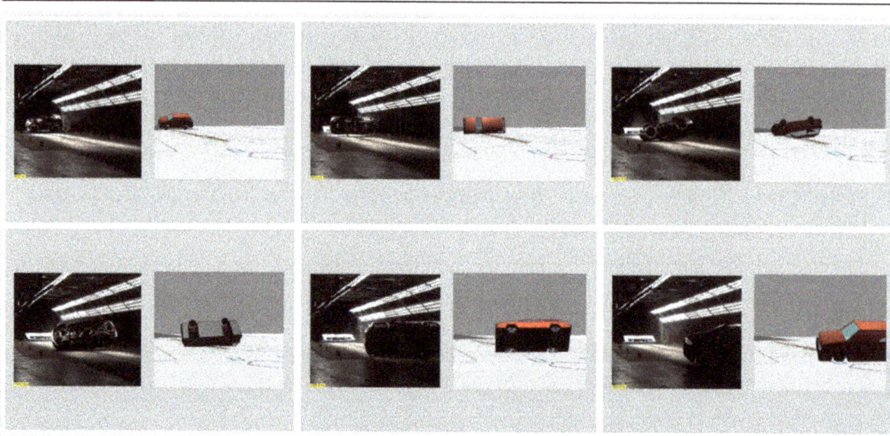

vehicles on a split-μ surface. This model could potentially be relevant to the analysis of the loss-of-control phase of a rollover. Zebala notes, "The accuracy of simulated vehicle movement in the phase of extreme braking is greatly affected by the ABS model. However, it is not practical to include the actual, individual algorithms used in different cars. This is why in [PC-Crash] universal algorithms are used whose task is to meet general criteria of ABS operation in order to solve the majority of typical problems that appear in road accident analysis." The PC-Crash ABS model controls the level of braking on a wheel-by-wheel basis in order to ensure that each individual wheel does not lock up. For some real-world ABS, the braking level for the rear wheels is adjusted together and is limited to the braking level appropriate for the tire on the lower coefficient of friction surface. Zebala refers to this as the "select-low principle," and he notes that the PC-Crash ABS model does not contain this logic. He reported good agreement between PC-Crash and the extreme braking tests on the split-μ surface when this logic was taken into account by manually reducing the braking level on the rear wheel that was on the high-μ surface.

Zebala [22] used PC-Crash Version 9.2 in conjunction with the linear tire model to analyze lane change maneuvers involving vehicles with reduced tire pressure. This study, too, could potentially be relevant to analyzing the loss-of-control phase of a rollover. Zebala stated, "The authors have made an attempt at parameterization of a tire model in PC-Crash program, based on the results of experimental research into vehicle motion. Attention was focused on one program and one, so-called bilinear tire model, which was selected because of its extreme simplicity from the point of view of an average user. The program validation process enabled first the tuning of the lateral force characteristics of undamaged tires, and next of tires of lowered pressure. Although good compatibility between the simulated and real results was reached, it should be emphasized that the selected maximum slip angles (describing tire lateral force characteristics) have to be treated with the maximum uncertainty of ±33% taken into account. This range also covers the influence of elements neglected in the modeling of the whole vehicle." Ultimately, Zebala's study sought reasonable inputs to obtain the best match with his test data, and so while in part a validation study, this study should also be categorized as part calibration study.

STUDIES THAT ASSUME THE VALIDITY OF PC-CRASH

Viano and Parenteau [23] reported that "a multi-disciplinary effort was conducted to define the most relevant tests that reflect real-world rollover crashes and injuries. That work prompted a series of rollover tests, including new procedures in which vehicle and occupant kinematics were studied. Also, each test was simulated in mathematical models to study other parameters and scenarios in rollover crashes." This team utilized and validated PC-Crash software for its mathematical modeling of the vehicle motion for the rollover tests. They used MADYMO to simulate the occupant motion. The article by Viano and Parenteau reports that PC-Crash and MADYMO simulations were used "to assure robust testing, sensing and algorithms. The mathematical models were applied to each specific test condition, validated and used for evaluation of parameters influencing rollover sensing requirements. The simulations were found to be robust representations of a vehicle rollover...Excellent comparability was demonstrated between the tests and the simulation."

Schubert [24] reported reconstructions of 3 rollover crashes. He used PC-Crash to simulate the events and to derive crash sensor signals for the events. He stated that "insights derived from these events can aid in development of robust rollover detection algorithms and calibrations."

Ootani and Pal [25, 26] of Nissan Motor Company published a pair of studies in 2007 in which they used PC-Crash to simulate real-world, soil-tripped rollovers from the NASS-CDS database. In these studies, they noted that "through a process of iteration the overall kinematics of the vehicle movement before and after the crash was captured.

The output of this PC-Crash simulation was then used as the initial input conditions (i.e., speed, deceleration, etc.) of a detailed finite element analysis."

Rose [27] used PC-Crash to explore the effects of high rates of yaw rotation and steering on the deceleration rates experienced by spinning vehicles. Rose noted that PC-Crash simulation is an efficient and effective manner in which to deal with these effects that are more difficult to incorporate analytically. He stated: "These software packages [HVE and PC-Crash] account for the actual tire travel distances and the effects of rotational kinetic energy using a vector mechanics approach and numerical integration."

Suspension Parameters for PC-Crash

Kiefer [28] used PC-Crash 8.2 to model 26 instrumented handling tests using 1998 and 1999 Ford Explorers. The handling tests involved rapid steering inputs at vehicle speeds between 30 and 60 mph [48.3 and 96.6 kph]. The vehicle weights, center-of-gravity positions, suspension stiffness parameters, tire parameters, and steering angle were measured and used as inputs into the simulations. The suspension stiffness coefficients were determined by measuring the change in the front and rear ride height of the test vehicles under different loading conditions. PC-Crash does not model the tire compliance separate from the suspension compression, and so the stiffness coefficients included the compliance of the tires. Kiefer noted that "the PC-Crash linear damping rate was increased to approximately twice the default value...The authors did not have shock curves or other suspension information to use as input for the model. As a result, the suspension damping values were increased from the default until the modeled response approached the test data." Kiefer concluded that "PC-Crash appeared to be a reasonable tool for modeling gross vehicle response. In addition, PC-Crash correctly predicted whether or not the test vehicle would experience rollover instability in a majority of the cases."

Heinrichs [29] described a method of measuring suspension properties for use in PC-Crash, and he reported suspension stiffness and damping values for 26 vehicles. The resulting values from Heinrichs' study are included below in English units (Table 9.1). The appendix to this chapter contains these values in SI units. When vehicle-specific data is not available for a particular vehicle, PC-Crash calculates default suspension stiffness values that are proportional to the static load at each wheel and a damping coefficient that is related to that stiffness. Heinrichs reported that "the deflection of the suspension under the vehicle's own static weight was lower than PC-Crash defaults. Average pitch deflection was 8 cm and average roll deflection was 6 cm, whereas the PC-Crash 'normal' and 'stiff' defaults are 15 cm and 10 cm, respectively." This means that the PC-Crash default suspension stiffness coefficients are generally too low.

In a 2012 study, Heinrichs [30] showed that the reconstructed speeds obtained with PC-Crash for a planar collision were "insensitive to four-fold variations in suspension parameters." One would expect the suspension parameters to matter more for rollover simulation than for simulating planar impacts, and Heinrichs' 2013 study bore this out. In that study, Heinrichs explored the sensitivity of PC-Crash rollover simulations to variations in the suspension and damping (he used PC-Crash 9.1). Specifically, Heinrichs ran simulations for each of the 26 vehicles he tested using both default and measured suspension parameters. He stated that the "simulations consisted of a steering maneuver to the right followed by a hard steering maneuver to the left...the lowest initial vehicle speed was found that resulted in a rollover. After the critical rollover speed was determined, new simulations were run using the measured suspension parameters. The initial speed for these simulations was a randomized value above the critical rollover speed. The results of these simulations were used as trajectory data; it was treated as field data

TABLE 9.1 Measured suspension parameters from Heinrichs [29]

Vehicle	Class	Weight (lb)	Height (in.)	k_{pitch} (lb/in.)	k_{roll} (lb/in.)	k_{roll}-k_{pitch} (lb/in.)	c (lb-s/ft)
2010 Subaru Impreza WRX	AWD	3212	58.1	887	1138	251	389
2001 Audi Allroad	AWD	4202	60.1	765	990	225	542
2006 Mazda 6 Speedsix	AWD	3638	56.7	777	1308	530	729
2010 Subaru Forester	AWD	3336	65.9	718	1135	417	397
2008 Infiniti G32xs	AWD	3874	57.2	866	1328	462	546
2010 Mazda 6 Sedan	FWD	3468	56.7	640	1002	362	493
2009 Volkswagen GTI	FWD	3168	58.4	752	1252	500	662
2004 Nissan Altima	FWD	3190	57.9	615	991	376	386
2004 Toyota Prius	FWD	2974	58.1	571	735	164	314
2000 Honda Civic	FWD	2339	54.1	570	501	-69	325
2011 Toyota Venza	FWD	4173	63.4	844	1245	402	712
2008 Mazda 3	FWD	2866	57.7	672	993	321	499
2001 Ford Focus	FWD	2659	56.3	540	751	210	274
2010 Honda Fit	FWD	2491	60.0	627	761	134	408
2011 Kia Forte	FWD	2899	55.1	800	1271	471	470
2007 Mitsubishi Outlander	FWD	3633	66.1	756	1318	562	555
2002 Chevrolet Cavalier 4dr	FWD	2676	53.0	649	581	-67	314
2006 Toyota Tundra (lifted)	Pickup	5426	70.5	1198	1344	145	528
2003 Infiniti G35	RWD	3602	57.7	795	1228	433	603
2009 Ford Escape	SUV	3371	67.9	719	1138	419	315
2008 Lexus RX400h	SUV	4612	66.4	873	1284	411	484
2012 Chevrolet Equinox	SUV	3768	66.3	922	1464	542	535
2001 Toyota RAV4	SUV	3020	65.7	673	1015	342	207
2010 Ford Escape	SUV	3680	67.9	743	1154	411	367
2006 Honda CRV	SUV	3530	66.2	897	1164	266	540
2010 Honda Odyssey	Van	4579	68.8	964	1268	304	478

for a reconstruction using the program default suspension values. For each vehicle, an initial vehicle speed was found that gave an operator-assessed best fit to the trajectory data. Vehicle speeds, steering, braking, and friction coefficients could be altered, but geometry, inertial and suspension parameters were fixed." Heinrichs found that "the critical speed required for a vehicle to roll in a PC-Crash simulation was about proportional to the suspension stiffness." He further found that the sensitivity of PC-Crash to variations in suspension stiffness for rollover reconstruction was minimized if evidence from all three phases of the rollover (loss of control, trip, and roll) was incorporated in the simulation.

In these findings, Heinrichs inherently distinguishes between two purposes for which PC-Crash could be used. For the purpose of predicting rollover, Heinrichs found that the results were sensitive to the suspension parameters. For the purpose of reconstructing a rollover – where the reconstructionist can incorporate evidence from all three phases of the rollover – the sensitivity to suspension parameters was minimized. One issue not addressed by Heinrichs is the fact that the vehicle suspension is subjected to drastically different loadings during each of the phases of the rollover, with the most significant loading occurring during the roll phase. The suspension and damping values reported by Heinrichs were obtained under loading conditions that would be representative of the loss-of-control phase, not the roll phase. Thus, these values may not work well when modeling the roll phase.

For the purpose of the discussion here, the authors compared the suspension stiffness and damping values that Heinrichs reported for each of the 26 vehicles in his study to the PC-Crash default calculated values for each of those vehicles. This comparison utilized the "normal" suspension settings in PC-Crash 10.0. For the front suspension, this comparison resulted in the conclusion that the measured suspension values were on average 4.5 times more stiff than the PC-Crash calculated suspension stiffness coefficients in pitch ($\sigma = 0.57$). Figure 9.3 shows a graphical comparison of the calculated and measured values for the front suspension. The linear fit displayed on this graph has been forced through the origin in order to relate the PC-Crash and measured values through a simple multiplier ($R^2 = 0.6367$). Another option for calculating realistic pitch suspension stiffness coefficients is presented graphically in Figure 9.4. This figure plots the relationship between the suspension stiffness in pitch and the vehicle weight. The linear fit displayed on this graph has a slightly better R^2 value than that of the previous graph (0.6967). This relationship could be used to calculate stiffness values for a particular vehicle.

Most of the vehicles tested by Heinrichs exhibited greater suspension stiffness in roll than in pitch. PC-Crash 10.0 allows the user to enter this additional roll stiffness. The measured additional suspension stiffness in roll was on average 3.8 times greater than the PC-Crash calculated value, but there was wide variability in the multiplier. In this instance, a linear fit had a very low R^2 value of 0.1364 (Figure 9.5). Increasing the suspension stiffness in roll has the effect of increasing the lateral forces necessary to generate the roll. Thus, in a reconstruction context, stiffer suspension would mean a

FIGURE 9.3 Comparison of calculated PC-rash front suspension pitch stiffness to measured.

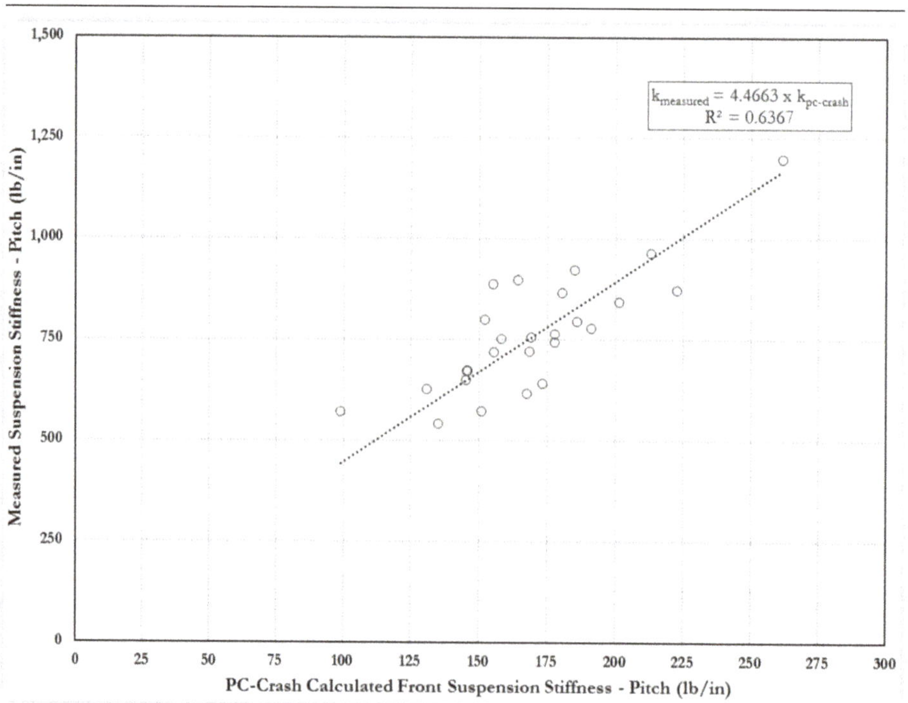

FIGURE 9.4 Relationship between vehicle weight and suspension stiffness in pitch.

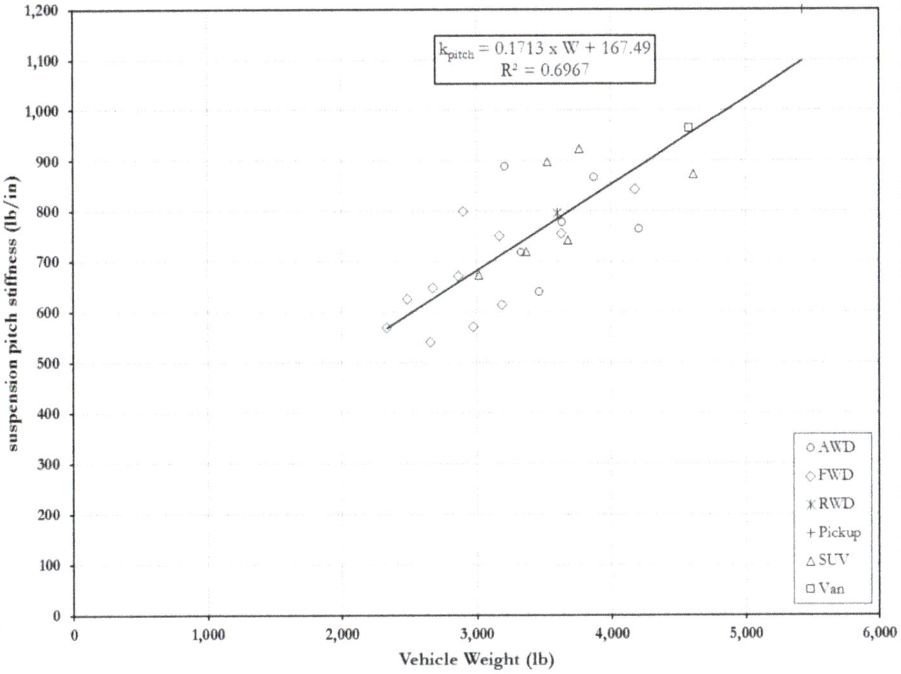

FIGURE 9.5 Comparison of default PC-Crash additional roll suspension stiffness to measured.

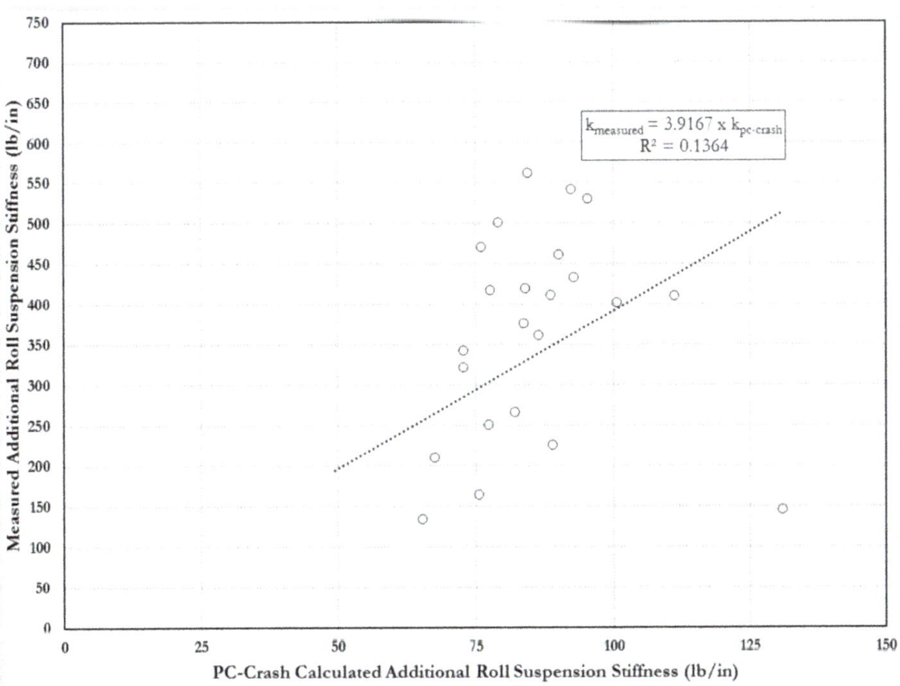

FIGURE 9.6 Comparison of default PC-Crash damping to measured.

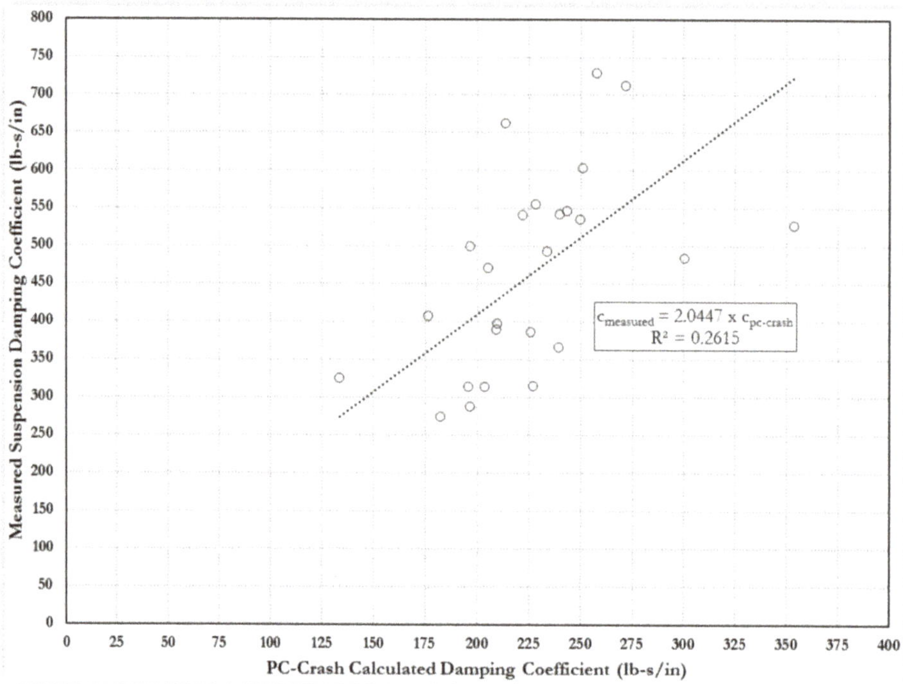

higher resulting speed. A conservative assumption would, therefore, be to leave the setting for additional roll stiffness at the PC-Crash default value.

The damping coefficients measured by Heinrichs were on average 2 times greater than the PC-Crash defaults; though again, in this instance, there was considerable scatter in the data, and a linear fit to the data had a low R^2 value of 0.2615 (Figure 9.6). That acknowledged, Kiefer [28] doubled the PC-Crash damping coefficient for his simulations with favorable results. For the purpose of rollover reconstruction, doubling the damping coefficient is likely an adequate assumption.

Simulation of Asay's Isuzu Rodeo Test

Previous chapters demonstrated rollover reconstruction techniques using the Isuzu Rodeo rollover test that Asay reported in 2010 (Test #2). The roll phase of this test was analyzed in Chapter 3 (constant deceleration rate approach) and Chapter 4 (linearly decreasing deceleration rate approach). The trip phase was analyzed in Chapter 5 and the loss of control in Chapter 6. This chapter continues the analysis of this test, in this instance using PC-Crash to reconstruct all three phases of the rollover.

In the test analyzed here, the Isuzu was towed up to and released at a speed of 73.5 mph (118.3 kph). After release, the vehicle was steered with a sharp left steering input of approximately ¼ turn, causing the vehicle to travel to the left across the roadway and to yaw counterclockwise, developing a slip angle of sufficient magnitude that tire marks were deposited on the road surface. Then, one second later, the first steering input was followed by a severe steering input back to the right. The vehicle continued off the left side of the road into the dirt, but reversed its yaw direction, developing a significant slip angle as it yawed in a clockwise manner. The vehicle deposited tire marks from the

FIGURE 9.7 Evidence diagram created for Asay's Isuzu Rodeo test.

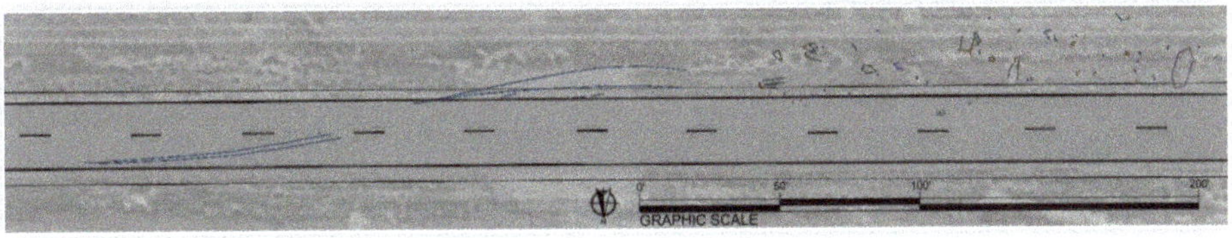

clockwise yaw and furrows in the dirt and then began rolling over. The vehicle rolled over 7 times.

The physical evidence diagram included below (Figure 9.7) was used as a background image for the simulation. During the process of producing a simulation of this test, the results were judged based on how closely the calculated vehicle motion followed the evidence depicted on this diagram. A three-dimensional terrain was also produced based on Asay's survey data and was used in the simulation. Manufacturer specifications provided the vehicle dimensions, and the vehicle weight reported by Asay was used (4,266 lb). Moments of inertia were estimated based on formulas in MacInnis [31] and Allen [32]. PC-Crash TM-Easy tire model was used with the default parameters. The suspension parameters were varied and the effects of these variations are discussed below. Representative shape parameters for the vehicle body of the Isuzu were measured from a computer model and entered for the simulation. A dxf model was used for display purposes, but not for generating the vehicle shape in the calculations.

The initial vehicle speed of 73.5 mph was entered, and then steering inputs were input in an attempt to match the physical evidence. Friction zones were used for the trip phase with the friction values being varied in order to induce roll of the vehicle at the correct location based on the tire marks.

Figure 9.8 shows the motion of the vehicle in the final optimized simulation. Figure 9.9 shows this same motion, but is zoomed in on the loss-of-control and trip phases. Initial simulation runs made it clear that the PC-Crash suspension parameters had a significant influence on both the roll distance and the number of rolls experienced by the Isuzu in the simulation. The vehicle body stiffness also had a significant influence on the number of rolls. To optimize the simulation, the suspension damping was set at twice the default value and then the suspension stiffness was varied to obtain the correct roll distance. The vehicle body stiffness was then varied to achieve the correct number of rolls in this distance. This proved to be a highly iterative process as the optimization neared completion.

FIGURE 9.8 Simulated motion overlaid on the evidence diagram.

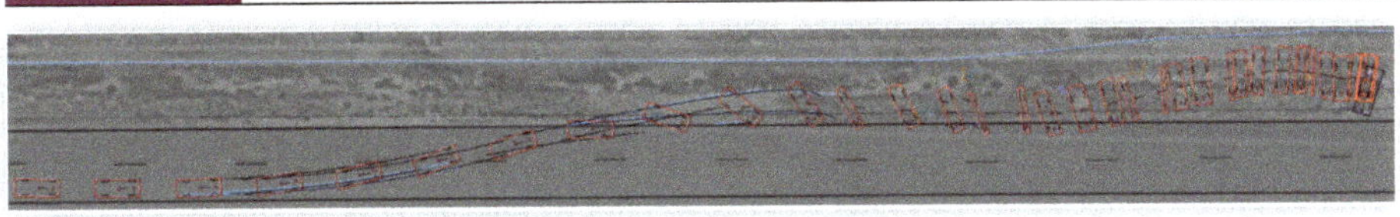

FIGURE 9.9 Simulated motion overlaid on evidence diagram - loss of control and trip.

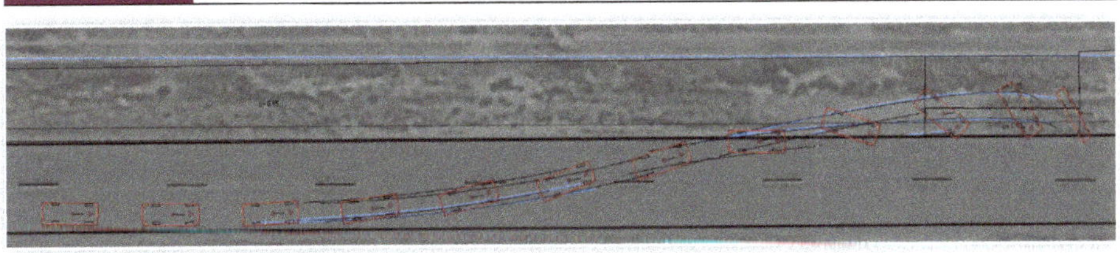

FIGURE 9.10 Comparison of simulated and actual roll velocities.

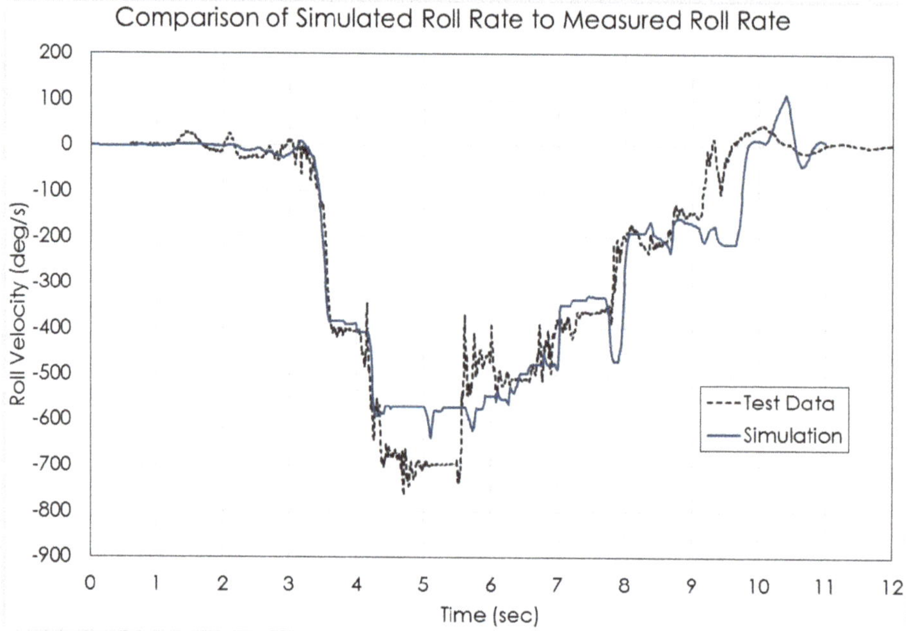

Ultimately, we obtained a simulation that exhibited an excellent visual match with the pre-roll tire marks and furrows. The vehicle in this simulation rolled 7 times, just as the vehicle did in the actual test. The rest position of the vehicle was also reasonably matched. The first half of the simulated roll trajectory was closer to the paved portion of the roadway than the actual roll trajectory. After that point, there was good agreement between the simulated and actual roll trajectories. Figure 9.10 is a graph showing the roll velocities of the vehicle during the roll phase in the simulation compared to those during the actual test. As this graph shows, there is a very close match until about 4¼ s into the simulation, where the roll velocities in the simulation are lower than the actual roll velocities. The peak roll velocity in the simulation was approximately 100°/s lower than the actual peak (650°/s versus 750°/s). In some areas near the peak, the roll velocity in the simulation is as much as 125°/s less than actual. This is compensated for by several later regions in the simulation where the roll velocities are greater than actual. The total roll duration in the simulation is approximately ½ s longer than actual.

References

1. Kleijnen, J.P.C., "Theory and Methodology: Verification and Validation of Simulation Models," *European Journal of Operational Research* 82 (1995): 145-162, doi:10.1016/0377-2217(94)00016-6.
2. Balci, O., "Verification, Validation and Accreditation of Simulation Models," *Proceedings of the Winter Simulation Conference*, Atlanta, GA, 1997.
3. Robinson, S., "Simulation Model Verification and Validation: Increasing the Users' Confidence," *Proceedings of the Winter Simulation Conference*, 1997.
4. Carson, J.S., "Model Verification and Validation," *Proceedings of the Winter Simulation Conference*, Stillwater, OK, 2002.

5. Sargent, R.G., "Verification and Validation of Simulation Models," *Proceedings of the 2003 Winter Simulation Conference*, New Orleans, LA.
6. Day, T. and Hargens, R.L., "Application and Misapplication of Computer Programs for Accident Reconstruction," SAE Technical Paper 890738, 1989, doi:10.4271/890738.
7. Wach, W., *Simulation of Vehicle Accidents Using PC-Crash*, (Linz, Austria, Institute of Forensic Research Publishers, October 23, 2012), ASIN: B009VJC5AO.
8. Rice, R.S., Dell'Amico, F., and Rasmussen, R.E., "Automobile Driver Characteristics and Capabilities-The Man-Off-The-Street," SAE Technical Paper 760777, 1976, doi:10.4271/760777.
9. Mazzae, E.N., Barickman, F., Baldwin, G.H.S., and Forkenbrock, G., "Driver Crash Avoidance Behavior with ABS in an Intersection Incursion Scenario on Dry Versus Wet Pavement," SAE Technical Paper 1999-01-1288, 1999, doi:10.4271/1999-01-1288.
10. Mazzae, E.N., Baldwin, G.H.S., and McGehee, D.V., "Driver Crash Avoidance Behavior with ABS in an Intersection Incursion Scenario on the Iowa Driving Simulator," SAE Technical Paper 1999-01-1290, 1999, doi:10.4271/1999-01-1290.
11. Carr, L., Liebbe, R., Crimeni, J., and Johnston, M., "Motor Vehicle Driver Characteristics-Crash Avoidance Behavior," SAE Technical Paper 2007-01-0449, 2007, doi:10.4271/2007-01-0449.
12. Bartlett, W., Masory, O., and Wright, B., "Driver Abilities in Closed Course Testing," SAE Technical Paper Number 2000-01-0179, 2000, doi:10.4271/2000-01-0179.
13. Forkenbrock, G., "An Assessment of Human Driver Steering Capability," DOT HS 809 875, National Highway Traffic Safety Administration, Washington, DC, June 2005.
14. Day, T. and Garvey, J., "Applications and Limitations of 3-Dimensional Vehicle Rollover Simulation," SAE Technical Paper 2000-01-0852, 2000, doi:10.4271/2000-01-0852.
15. Steffan, H. and Moser, A., "The Collision and Trajectory Models of PC-CRASH," SAE Technical Paper 960886, 1996, doi:10.4271/960886.
16. Steffan, H. and Moser, A., "How to Use PC-CRASH to Simulate Rollover Crashes," SAE Technical Paper 2004-01-0341, 2004, doi:10.4271/2004-01-0341.
17. Rose, N. and Beauchamp, G., "Analysis of a Dolly Rollover with PC-Crash," SAE Technical Paper 2009-01-0822, 2009, doi:10.4271/2009-01-0822.
18. Rose, N., Neale, W., Fenton, S., Hessel, D. et al., "A Method to Quantify Vehicle Dynamics and Deformation for Vehicle Rollover Tests Using Camera-Matching Video Analysis," *SAE Int. J. Passeng. Cars - Mech. Syst.* 1, no. 1 (2009): 301-317, doi:10.4271/2008-01-0350.
19. Rose, N. and Beauchamp, G., "Development of a Variable Deceleration Rate Approach to Rollover Crash Reconstruction," *SAE Int. J. Passeng. Cars - Mech. Syst.* 2, no. 1 (2009): 308-332, doi:10.4271/2009-01-0093.
20. Andrews, S.B., Partain, P.M., Penfroe, D.A., and Gilbert, M., "A Comparison of Computer Modeling to Actual Data and Video of a Staged Rollover Collision," *Enhanced Safety of Vehicles (ESV) Conference*, Stuttgart, Germany, 2009, Paper Number 09-0346.
21. Zebala, J., Wach, W., Ciepka, P., Janczur, R. et al., "Verification of ABS Models Applied in Programs for Road Accident Simulation," *SAE Int. J. Passeng. Cars - Mech. Syst.* 3, no. 1 (2010): 72-99, doi:10.4271/2010-01-0070.

22. Zebala, J. and Wach, W., "Lane Change Maneuver Driving a Car with Reduced Tire Pressure," SAE Technical Paper 2014-01-0466, 2014, doi:10.4271/2014-01-0466.
23. Viano, D.C. and Parenteau, C.S., "Rollover Crash Sensing and Safety Overview," SAE Technical Paper 2004-01-0342, 2004, doi:10.4271/2004-01-0342.
24. Schubert, P., "Real-World Rollovers Reconstructed from Interviews and Measurements," SAE Technical Paper 2006-01-0060, 2006, doi:10.4271/2006-01-0060.
25. Ootani, R. and Pal, C., "Effective Numerical Simulation Tool for Real-World Rollover Accidents by Combining PC-Crash and FEA," SAE Technical Paper 2007-01-1773, 2007, doi:10.4271/2007-01-1773.
26. Ootani, R. and Pal, C., "Soil Trip Rollover Simulation and Occupant Kinematics in Real World Accident," SAE Technical Paper 2007-01-3680, 2007, doi:10.4271/2007-01-3680.
27. Rose, N., Carter, N., Moser, A., and Beauchamp, G., "Post-Impact Dynamics for Vehicles with a High Yaw Velocity," SAE Technical Paper 2016-01-1470, 2016, https://doi.org/10.4271/2016-01-1470.
28. Kiefer, A., Bilek, D., Moser, A., and Webb, A., "A Comparison Study between PC-Crash Simulation and Instrumented Handling Maneuvers," SAE Technical Paper 2011-01-1121, 2011, doi:10.4271/2011-01-1121.
29. Heinrichs, B., Goulet, J., Fix, R., and King, M., "Measuring and Modeling Suspensions of Passenger Vehicles," SAE Technical Paper 2013-01-0774, 2013, doi:10.4271/2013-01-0774.
30. Heinrichs, B., Mac Giolla Ri, B., and Hunter, R., "Sensitivity of Collision Simulation Results to Initial Assumptions," *SAE Int. J. Passeng. Cars - Mech. Syst.* 5, no. 2 (2012): 807-832, doi:10.4271/2012-01-0604.
31. MacInnis, D., Cliff, W., and Ising, K., "A Comparison of Moment of Inertia Estimation Techniques for Vehicle Dynamics Simulation," SAE Technical Paper 970951, 1997, doi:10.4271/970951.
32. Allen, R., Klyde, D., Rosenthal, T., and Smith, D., "Estimation of Passenger Vehicle Inertial Properties and Their Effect on Stability and Handling," SAE Technical Paper 2003-01-0966, 2003, doi:10.4271/2003-01-0966.

Appendix

TABLE 9.2 Measured suspension parameters from Heinrichs [29], SI units

Vehicle	Class	Mass (kg)	Height (cm)	k_{pitch} (N/m)	k_{roll} (N/m)	c (N-s/m)
2010 Subaru Impreza WRX	AWD	1457	147.6	155383	199257	5681
2001 Audi Allroad	AWD	1906	152.7	133924	173350	7912
2006 Mazda 6 Speedsix	AWD	1650	144.0	136150	229052	10643
2010 Subaru Forester	AWD	1513	167.4	125667	198747	5796
2008 Infiniti G32xs	AWD	1757	145.3	151735	232578	7968
2010 Mazda 6 Sedan	FWD	1573	144.0	112119	175431	7200
2009 Volkswagen GTI	FWD	1437	148.3	131634	219258	9664
2004 Nissan Altima	FWD	1447	147.1	107707	173495	5639
2004 Toyota Prius	FWD	1349	147.6	100040	128754	4583
2000 Honda Civic	FWD	1061	137.4	99838	87816	4750
2011 Toyota Venza	FWD	1893	161.0	147722	218096	10396
2008 Mazda 3	FWD	1300	146.6	117615	173874	7286
2001 Ford Focus	FWD	1206	143.0	94649	131449	3992
2010 Honda Fit	FWD	1130	152.4	109779	133236	5949
2011 Kia Forte	FWD	1315	140.0	140149	222591	6864
2007 Mitsubishi Outlander	FWD	1648	167.9	132348	230834	8103
2002 Chevrolet Cavalier 4dr	FWD	1214	134.6	113582	101784	4578
2006 Toyota Tundra (lifted)	Pickup	2461	179.1	209888	235345	7703
2003 Infiniti G35	RWD	1634	146.6	139308	215077	8806
2009 Ford Escape	SUV	1529	172.5	125924	199372	4599
2008 Lexus RX400h	SUV	2092	168.7	152926	224825	7066
2012 Chevrolet Equinox	SUV	1709	168.4	161525	256417	7811
2001 Toyota RAV4	SUV	1370	166.9	117799	177682	4195
2010 Ford Escape	SUV	1669	172.5	130128	202154	5352
2006 Honda CRV	SUV	1601	168.1	157144	203815	7887
2010 Honda Odyssey	Van	2077	174.8	168810	222034	6978

CHAPTER 10

Event Data Recorders in Rollover Reconstruction

The Crash Data Retrieval (CDR) system was released in 2000 [1]. This system provided crash reconstructionists access to crash-related data recorded by onboard event data recorders (EDRs), initially from some 1994 model year and newer General Motors (GM) vehicles. Beginning with model year 1999, the data that could be accessed on some GM vehicles included pre-crash data. In 2003, some 2001 and newer Ford vehicles were added to the list of vehicles supported by the CDR system. In August of 2006, the National Highway Traffic Safety Administration (NHTSA) issued Rule 49 of the Code of Federal Regulations (CFR) Part 563. This rule aimed to standardize the data and retrieval methods for light vehicles in which the manufacturers were already voluntarily recording event data [2]. This resulted in additional manufacturers partnering with the CDR system to make their onboard data available to accident reconstructionists.

In recent years, the data recorded by the onboard vehicle systems and accessible with the CDR system has become more voluminous. Newer electronic stability control (ESC) systems, for instance, monitor the wheel speeds, the vehicle yaw rate, and the steering wheel position, and this data is now included in the CDR reports for some vehicles. Similarly, side curtain airbag systems monitor the roll rate of the vehicle, and this data is sometimes available via the CDR system.

One might imagine that the increasing amount of data accessible from the vehicle would result in a diminished role for accident reconstructionists. After all, some vehicles now report much of the data that an accident reconstructionist would have acquired in the past through classic reconstruction techniques. However, there are many instances when an accident reconstructionists is needed to interpret the data. Oftentimes, the correct interpretation of the data from the vehicle is not apparent on the surface, and someone needs to analyze its meaning within the context of the crash. The early CDR reports were short, often with a page or two of data, and were relatively easy to digest. Now, it is not uncommon for CDR reports to be over 60 pages long. In many cases, this data is not laid out in a way that enables a lay person to read the report and understand what happened in the crash. Without a technical background in accident reconstruction, it can be difficult to know which recorded data is related to the subject crash, which data is useful, what limitations exist in the data, what specifically was happening to the vehicle when the data was collected, and where and when during the sequence specific data was collected. Some of the data from the vehicle may not be understandable without a full accident reconstruction. Thus, while the types of evidence available have shifted, reconstructionists are still often called upon to analyze a crash and present the results in a clear way.

Vehicle EDRs often report data pertinent to each phase of a rollover crash–the loss of control, the trip, and the roll. The types of EDR data relevant to each of these phases are discussed below. The accuracy, limitations, and use of each of these types of data are also discussed.

EDR Data: Loss-of-Control Phase

The loss-of-control phase typically begins with the vehicle in a steady-state condition, traveling straight down the road at a relatively constant speed. In response to driver steering inputs, the vehicle then develops a sideslip angle. The loss-of-control phase ends when the trailing side tires begin to lift off the ground. Thus, for the loss-of-control phase, the accuracy of EDR-reported speeds under steady-state and sideslip conditions are of interest. The effect of brake application on EDR-reported speed accuracy is also relevant.

Speed Reported by Passenger Car EDRs

In 2016, Bortles, Biever, Carter, and Smith presented a comprehensive review of the literature related to pre-crash speeds reported by original equipment EDRs installed in passenger vehicles. They reported that "event data recorders accurately measure and record the vehicle wheel (or transmission output) speed and integrated accelerations of the module. Reported values of vehicle velocity change (ΔV) and Pre-Crash vehicle speed tend to be less than the actual values. Numerous factors may contribute to the under-reporting of this data. For Pre-Crash vehicle speed, the predominant factors include longitudinal wheel slip and sideslip…Analysts should consider event recorder data within the context of an accident reconstruction and account for factors that cause discrepancies between the reported and actual values."

In reaching these conclusions, Bortles and his colleagues compiled 187 sources of information, including peer-reviewed studies, textbooks, legal opinions, governmental rulemaking policies, industry publications, and presentations pertaining to EDRs. Sixty-four of those studies contained physical test data related to the accuracy of the EDR data. Of those 64 studies, 27 of them "contained paired data points from EDR and independent instrumentation suitable to validate the accuracy of ΔV and Pre-Crash vehicle speed… Of these 27 papers, there were nine that contained Pre-Crash data, nine that contained ΔV data and nine that contained both data types."

The speeds reported by passenger vehicle EDRs are typically measured by sensors monitoring the output of the transmission or an average of the speed of the drive wheels.

While these sensors accurately report the speed they are measuring, this speed does not necessarily correspond to the actual over-the-ground speed of the vehicle. For example, during the loss-of-control phase of a rollover crash, the vehicle will often develop a sideslip angle, where there is a discrepancy between the heading of the vehicle and the velocity direction. When this occurs, the EDR-reported speed will be the approximate longitudinal component of the speed, but not the total over-the-ground speed. Also, if the driver applies the brakes or the accelerator, longitudinal wheel slip may be induced that will cause a difference between the wheel speed and the actual over-the-ground speed of the vehicle. Other possibilities include differences between the rolling radius of the OEM tires for a vehicle and the tires that are installed on the vehicle at the time of the crash and changes to the final drive ratio compared to the vehicle's original equipment. In addition to these causes, differences can also exist due to rounding, truncation, or unit conversion.

Bortles and his colleagues produced the graph of Figure 10.1 based on the studies they reviewed. In this graph, the actual pre-crash speed is plotted on the horizontal axis and the vertical axis plots the difference between the EDR-measured pre-crash speed and the actual pre-crash speed based on other instrumentation. Positive values on the vertical axis represent the EDR overestimating the actual speed and negative values represent the EDR underestimating the actual speed. The data in the graph has been sorted by vehicle operational condition. Steady-state driving (designated with green circles) was found to be associated with minor speed differences and a tendency to underreport vehicle speed. Active braking was associated with greater underreporting and more variance in reported vehicle speed.

Braking causes a discrepancy between the actual speed of the vehicle and the speed reported by the CDR, as shown in Figure 10.1. Broadly speaking, anything that affects

FIGURE 10.1 Differences between EDR-reported and actual pre-crash speeds in the studies reviewed by Bortles [3].

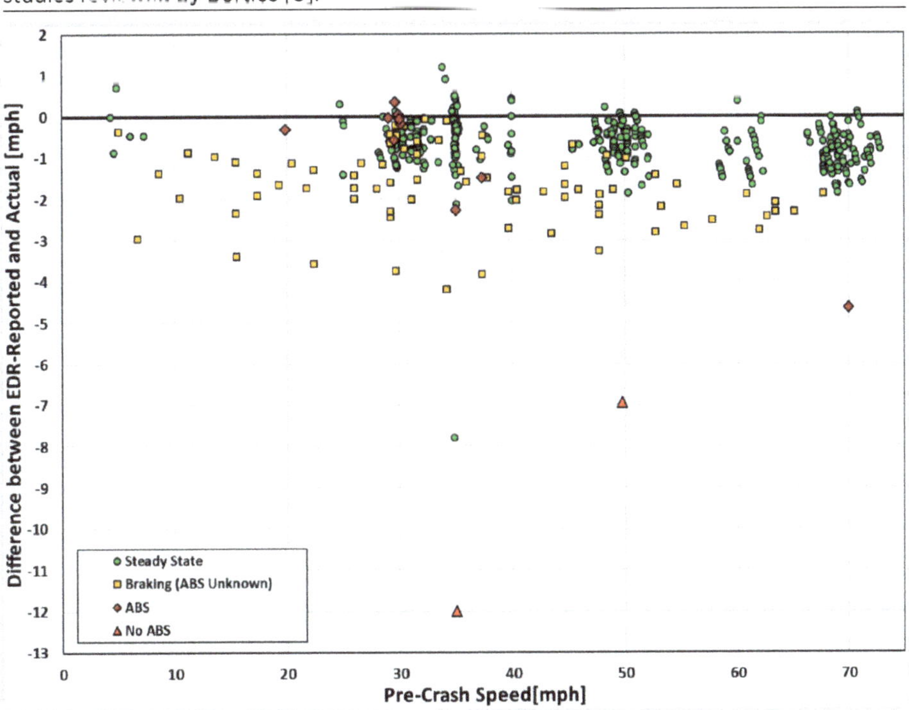

the rotational speed of the wheels can affect the speed reported by the CDR. When a vehicle yaws, and the tires generate a sideslip angle, the wheel speed will not be equal to the over-the-ground speed of the vehicle. As an extreme example, if a vehicle is sliding perpendicular to its heading (a slip angle of 90°), the wheels will stop rotating. In this case, the CDR could report a speed of zero when the speed could be significantly greater than zero. By measuring the wheel speeds or output speed of the transmission, the CDR is measuring the longitudinal speed of the vehicle. If the wheel speeds of the front tires are used by the EDR to measure speed, then steering can also affect the speed measurement along with vehicle yaw.

Several studies that are among those reviewed by Bortles et al. are worth mentioning specifically. Lawrence [4] reported that the pre-crash speeds reported by GM airbag sensing and diagnostic modules (SDMs) were within a 1.5 kph overestimate at low speeds and a 3.7 kph underestimate at high speeds. Lawrence established this range by artificially triggering the SDMs while the test vehicles were driven at a constant speed between 1 and 150 kph. The following vehicles were tested: a 2002 Pontiac Sunfire, a 2002 Chevrolet Malibu, and a 2002 Chevrolet Impala. The SDM-reported speeds were compared to the speeds measured by a calibrated 5th wheel.

Reust has reported that the accuracy of the speed data from Ford powertrain control module (PCM) EDRs "during low and maximum acceleration and during steady speeds is approximately ±0.8 mile per hour...During maximum brake application of a vehicle with an ABS system, the PCM-reported speed is underreported by approximately 0.5 to 14.3 percent at PCM-reported speeds of 20 miles per hour and higher...when a vehicle is in a critical speed yaw the PCM-reported speed will be underreported by approximately 0.02 to 2.9% when only steering has been used and will be underreported by approximately 0.11 to 5.2 percent when both steering and braking have been used" [5]. In 2007 and 2008, Reust reported similar testing for the speeds reported by the SDMs on GM vehicles [6, 7]. These units can record pre-event speeds at one-second intervals. Reust notes that newer SDMs also provide data related to lateral acceleration, yaw rate, activation of the ABS, and electronic stability control. Reust found that "based on the tests conducted for this paper, the accuracy of the SDM-reported speed during low and normal acceleration is approximately ±1.5 mile per hour. During maximum acceleration at full throttle, the accuracy is approximately ±1.5 mile per hour except during initial takeoff and manual gear shift. During coasting, the accuracy is approximately +0.5 to −2.0 miles per hour. During maximum brake application and when the ABS system is activated, the SDM-reported speed is underreported by approximately 8 to 18 percent at SDM-reported speeds of 30 miles per hour and higher. The exception was the pickup with an error of 25% which would occur briefly at approximately 0.28 seconds after brake application. When a vehicle is in a yaw the SDM-reported speed will be underreported by approximately 2 to 7% when only steering has been used and will be underreported by approximately 3 to 18% when both steering and braking have been used at SDM-reported speeds of 30 miles per hour or higher."

Ruth has reported the accuracy of the speed data from Ford PCM EDRs under steady-state conditions as within 1% [8]. In the same study, he examined the effect of low tire pressure and tire size. This examination resulted in the conclusion that a drop in pressure from 35 to 20 psi did not have a significant effect on the accuracy of the reported speed. A more significant effect was produced by putting an oversized tire on the vehicle. Ruth noted that "underreporting was expected and consistent with the change in tire diameter." He has also observed underreporting of the speed by PCM EDRs on Ford and Chrysler vehicles under heavy straight-line braking [9, 10].

Brown [11] evaluated the accuracy of pre-crash data reported by Toyota EDRs. He reported that "a total of thirty-three tests were performed on three vehicles. The pre-crash data and other parameters recorded by the EDR were compared to the vehicles' HS-CAN data or actual physical observation. For the pre-crash data, tolerance bars were determined for the Toyota EDR data based on the parameter refresh rate, recording truncation and data logger recording rate. For the three vehicle models tested, the Toyota EDR pre-crash data and other parameters were accurate when compared with the HS-CAN data or observations."

Effects of Sideslip

At the beginning of a loss-of-control sequence, before any braking by the driver or yawing of the vehicle, the vehicle is typically near steady state. Therefore, if the EDR captures a speed prior to the loss of control, then this speed is likely accurate, consistent with the results in Figure 10.1. As the crash sequence develops and the vehicle starts to yaw, the over-the-ground speed and CDR-reported speed are expected to diverge. By computing the longitudinal speed of the vehicle at various positions in an accident reconstruction, the CDR-reported speed and the reconstructed speed can be directly compared to one another. Thus, theoretically, the speed reported by the EDR would be related to the ground speed with the following equation:

$$S_{reported} = S_{ground} \cdot \cos \beta \tag{10.1}$$

In this equation, β is the vehicle sideslip angle.

In 2010, Ruth [14] reported an additional study examining the accuracy of an EDR on the PCM of a 2009 Ford Crown Victoria Police Interceptor (a rear-wheel drive vehicle with antilock brakes) during yaw rotations on low friction surfaces. The PCM EDR on this vehicle records and reports 25.2 s of vehicle operation at 0.2 s intervals. The EDR reports the vehicle speed, as measured at the transmission output shaft sensor, the brake status (on or off), the accelerator pedal position, and the throttle position.

This testing was conducted on a flooded skid pad at the Michigan State Police training facility in Lansing, Michigan. The coefficient of friction of the skid pad surface was approximately 0.18 when wet. For each test, the vehicle was accelerated to a speed of approximately 50 mph (80 kph), then the vehicle steered to the left and then hard to the right. Ruth reported that the test vehicle was "deliberately put into rotation and allowed to rotate to rest under three different conditions: heavy throttle applied initially, heavy braking applied and held, and light to no throttle applied." Six tests were run in each of these conditions. The tests resulted in vehicle yaw rotations between 103 and 486°.

In explaining the motivation for his research, Ruth noted that "rain may cause water puddling and hydroplaning resulting in loss of control and rotation. In snow and ice conditions, vehicles can also go into rotation. When pre-crash EDR data is recovered from these events, the reported speeds can change rapidly, seemingly violating laws of physics limiting acceleration of the complete vehicle mass. In reality, it is the wheel speed that is changing rapidly, and not the ground speed."

Data from the EDR was compared to data collected with a VBOX III data acquisition system. Ruth concluded that "under heavy throttle the tires break loose and spin up resulting in the PCM over-reporting ground speed regardless of slip angle." For the heavy braking condition, "the PCM under reports ground speed consistently, and as the rotation approaches 90 degrees and the wheels are scrubbing sideways and not rolling, the reported speed drops to or near zero...In half the runs the PCM EDR speed was at or above the longitudinal ground speed, and in half it was nearly equal or below it. The key factor identified was the time from when rotation began, or the time when brakes were applied, to when the vehicle achieved its first 90 degrees of rotation...in the runs where the EDR was reporting higher than expected values, the vehicle was slowing the wheels down faster trying to get them to the expected values–but just could not get them there fast enough." For the light-to-no-throttle condition, "the speed reported shows partial correlation to the slip angle cosine. In some cases as the vehicle rotation continues past 90 to 180 or 270 to 360 degrees, the tires begin rolling again and the EDR reported speed recovers to near the actual speed."

Analysis of EDR Data from the Loss-of-Control Phase

Even when EDR data is available, it may be necessary or helpful to reconstruct a rollover crash using the methods presented earlier in this book. Once the crash is reconstructed with those techniques, the reconstruction results can be compared with the speeds reported by the EDR. In many crash types, frontal collisions for example, it is often obvious where in the sequence the EDR was triggered to record data. This triggering time is known as Algorithm Enable (AE) and both airbag deployment and nondeployment events can be recorded. If a vehicle were to be involved in a single high-speed frontal impact to a tree, and the air bags deployed, AE most likely occurred during the tree impact. If the location of the tree was known, then aligning the EDR data to the reconstruction data is somewhat trivial. However, during a high-speed rollover event, there can be many vehicle-to-ground impacts that could trigger the EDR, so it is often unclear where specifically in the rollover sequence AE occurred. There could also be impacts during the loss-of-control or trip phase that produces sufficient accelerations to trigger an event. When EDR data is available following a rollover crash, there are often two questions that need answering:

Is the EDR data related to the subject rollover crash?

Where in the crash sequence was the various data recorded?

The following steps can be followed in most cases to answer these questions. This procedure can be carried out in Microsoft Excel, or another similar program.

1. Reconstruct the speed of the vehicle, starting from the rest position and working back to the start of physical evidence. This process will typically yield speeds at several positions defined by the physical evidence, including the speed at beginning of the roll phase, the speed at beginning of the trip phase, and several positions during the loss-of-control phase.
2. If striations are present in the tire mark evidence from the loss-of-control phase, quantify the driver's braking inputs and incorporate these into the speed analysis.
3. Calculate the longitudinal speed of the vehicle at each pre-roll position using Equation (10.1).
4. Plot the longitudinal speed of the vehicle versus time for the pre-roll positions.
5. Verify that the vehicle was equipped with OEM-sized tires. If the tires on the vehicle at the time of the crash are different than OEM sized, correct the EDR-reported speeds for the actual tire size.

6. Plot the longitudinal speeds from the EDR on the same graph as the reconstructed speeds.
7. Shift the time of the reconstructed longitudinal speed data or EDR-reported speeds until there is a reasonable alignment between the two.
8. If necessary, incorporate braking into the reconstructed speeds to achieve reasonable agreement between the reconstructed and EDR-reported speeds, in accordance with braking reported by the EDR.

Figure 10.2 depicts an example of a graph produced during reconstruction of an actual crash, comparing reconstructed longitudinal speeds (red) to EDR-reported speed (blue). In this case, the vehicle experienced a tire disablement at approximately −4 s. The tire disablement created tire slip, resulting in the irregular speed recording for approximately 1 s. The effect of tire disablements on EDR-reported speed is discussed further by Koch, Beauchamp, and Pentecost in their 2017 publication [13]. The driver lost control of the vehicle and the vehicle yawed off the side of the road where it rolled over. The EDR-reported speed dropped off to approximately 19 mph, the speed data then leveled off for 6/10th of a second before continuing to drop. The leveling of the speed data resulted when the vehicle became airborne. The driver was not applying brakes or throttle so the tires continued to rotate at a constant speed until the tires contacted the ground for the first time during the rollover. The reported speed then abruptly dropped. The time of the EDR-reported speed was shifted so that the rollover began at time zero.

The reconstructed longitudinal speed was also shifted, so the start of the rollover occurred at time zero. The EDR-reported speed of 19 mph was consistent with the rollover speed when adjusted for sideslip angle. Aligning the beginning of the rollover in the reconstruction with the beginning of the rollover observed in the data resulted in good

FIGURE 10.2 Comparison between reconstructed longitudinal speed and EDR-reported speed.

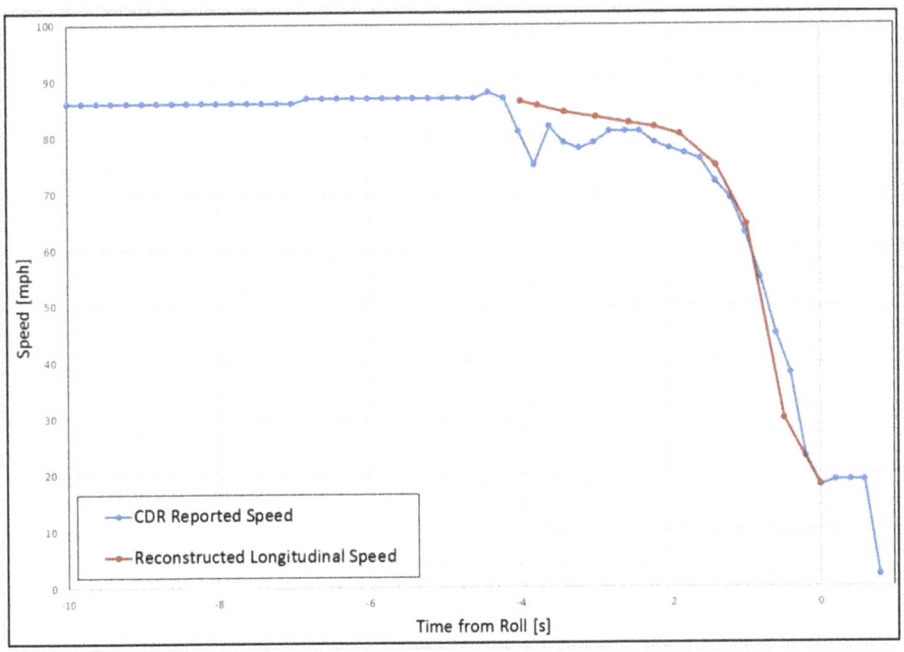

alignment with the remainder of the data. The reconstructed speed of the vehicle at the beginning of the physical evidence was in good agreement with the steady-state speed reported by the EDR.

Once the reconstruction and the EDR data are aligned in this way, the CDR data is effectively synced with the physical evidence. The analysist can then determine where along the path of the vehicle any point of interest within the CDR report occurred. For example, the location and conditions of the vehicle when the driver braked or when an airbag deployed can be examined. The accident reconstruction will often give context to the EDR data, while the data in the CDR report assists in refining and improving the accident reconstruction. The EDR data can also give additional data from before the vehicle started depositing physical evidence. In Figure 10.2 for example, the vehicle's speed and other data was reported approximately 6 s before the start of the physical evidence.

EDR Data: Trip and Roll Phases

The trip phase begins when the trailing side tires begin to lift off the ground and ends when the leading side tires leave the ground and the vehicle goes airborne. Some EDRs have the ability to record vehicle roll velocity. This data can be useful in analyzing the trip phase.

Figure 10.3 is the roll rate data recorded during Asay's Test #2 of the Isuzu Rodeo [14]. The roll rate stays relatively low until the beginning of the trip phase at approximately 2.5 s. The trailing tires then begin to lift and the roll rate increases. The vehicle goes airborne and begins to roll over at approximately 3 s at the end of the tire mark evidence. When the vehicle goes airborne, the forces that were acting to increase the roll rate vanish resulting in the roll rate plateauing for approximately ½ s. The vehicle then

FIGURE 10.3 Rollover data from the Isuzu Rodeo test (Asay [14], Test #2).

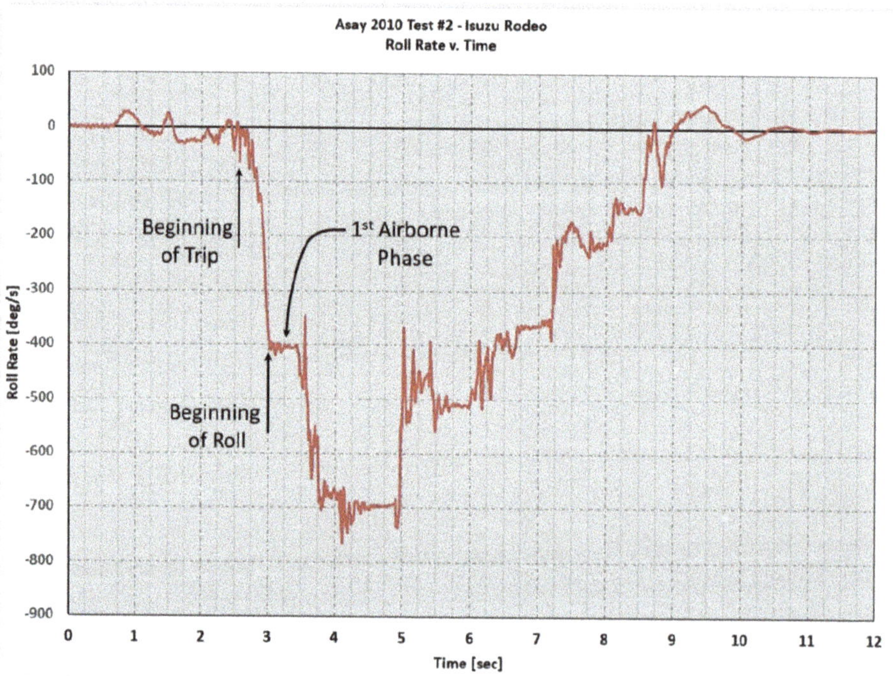

FIGURE 10.4 Roll rate data from a CDR report.

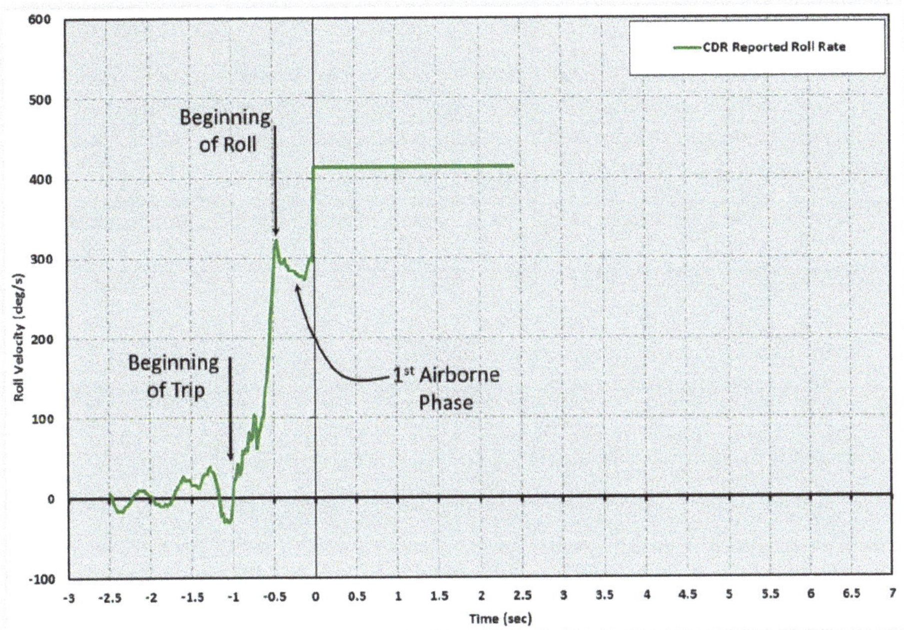

contacts the ground again and the roll rate continues to increase. The beginning of the trip phase, beginning of the roll phase, and first airborne phase are all identifiable in the data, and identified in Figure 10.3.

The pattern observed in Figure 10.3 is typical during rollover events, and this pattern is often visible in the EDR-recorded roll rates. Figure 10.4 depicts roll rate data from a CDR report. The beginning of the trip phase, beginning of the rollover phase, and first airborne phase are again indicated. At zero seconds, the roll rate sensor was maxed out.

Typically, the roll rate data and other EDR-reported data such as speed are synced in time. Therefore, identifying where in time the trip phase begins and ends allows the analyst to align the remainder of the CDR data with the reconstruction. In the case of the data in Figure 10.4, the beginning of the roll phase, or end of the tire marks in the reconstruction, corresponds to a time of −0.5 s in the CDR report.

References

1. Haight, R., "An Abbreviated History of CDR Technology," *Collision: The International Compendium of Crash Research*, 5, no. 1 (2010): 50-57, 64-73, ISSN 1934-8681.
2. National Archives and Records Administration, "Part 563 - Event Data Recorders," *Federal Register* 71, no. 166 (August 28, 2006): 51043-51048, Docket No NHTSA-2006-25666.
3. Bortles, W., Biever, W., Carter, N., and Smith, C., "A Compendium of Passenger Vehicle Event Data Recorder Literature and Analysis of Validation Studies," SAE Technical Paper 2016-01-1497, 2016, doi:10.4271/2016-01-1497.
4. Lawrence, J., Wilkinson, C., Heinrichs, B., and Siegmund, G., "The Accuracy of Pre-Crash Speed Captured by Event Data Recorders," SAE Technical Paper 2003-01-0889, 2003, doi:10.4271/2003-01-0889.

5. Reust, T., Morgan, J., and Ruth, R., "The Accuracy of Speed Recorded by a Ford PCM and the Effects of Brake, Yaw and Other Factors," *Collision Magazine* 3, no. 1 (2008): 48-59, ISSN 1934-8681.
6. Reust, T. and Morgan, J., "The Accuracy of Speed Recorded by an SDM and the Effects of Brake, Yaw and Other Factors," *Collision Magazine* 1, no. 1 (2006): 9-15.
7. Reust, T. and Morgan, J., "Detailed Comparison of Vehicle Speed and the Speed Recorded by an SDM," *Collision Magazine* 2, no. 2 (2007): 32-40.
8. Ruth, R., West, O., Engle, J., and Reust, T., "Accuracy of Powertrain Control Module (PCM) Event Data Recorders," SAE Technical Paper 2008-01-0162, 2008, doi:10.4271/2008-01-0162.
9. Ruth, R. and Reust, T., "Dynamic Accuracy of Powertrain Control Module (PCM) Event Data Recording During ABS Braking," *Collision Magazine*, 3, no. 1, Spring 2008, ISSN 1934-8681.
10. Ruth, R. and Reust, T., "Accuracy of Selected 2008 Chrysler ACM EDR's during Braking," *Collision Magazine*, 4, no. 1, Spring 2009.
11. Brown, R., Lewis, L., Hare, B., Jakstis, M. et al., "Confirmation of Toyota EDR Pre-crash Data," SAE Technical Paper 2012-01-0998, 2012, doi:10.4271/2012-01-0998.
12. Ruth, R., Brown, T., and Lau, J., "Accuracy of EDR During Rotation on Low Friction Surfaces," SAE Technical Paper 2010-01-1001, 2010, doi:10.4271/2010-01-1001.
13. Koch, D., Beauchamp, G., and Pentecost, D., "Deceleration Rates of Vehicles with Disabled Tires," SAE Technical Paper 2017-01-1427, 2017, doi:https://doi.org/10.4271/2017-01-1427.
14. Asay, A. and Woolley, R., "Rollover Testing of Sport Utility Vehicles (SUVs) on an Actual Highway," SAE Technical Paper 2010-01-0521, 2010, doi:https://doi.org/10.4271/2010-01-0521.

11

CHAPTER

Analyzing Occupant Ejections for Rollover Crashes

The Motion of Ejected Occupants

Reconstructionists are sometimes asked to determine when, and from where (i.e., out of which opening or portal), an occupant was ejected during a rollover. The accident reconstruction literature contains a number of models for projected people - struck pedestrians, bicyclists, and ejected motorcyclists. For instance, Collins [1] reported the following equation for calculating the launch speed of a person first undergoing an airborne trajectory and then landing and sliding to rest. This equation assumed a horizontal launch (no launch angle). In this equation, v_p is the throw speed of the person, h is the height of the person's center of gravity prior to the collision, g is the gravitational acceleration, μ is the coefficient of friction between the person and the road surface, and d_t is the throw distance (the total distance from the beginning of the projection to rest):

$$d_t = v_{proj}\sqrt{\frac{2h}{g}} + \frac{v_{proj}^2}{2\mu g} \tag{11.1}$$

Collins noted that there would be some speed loss from air resistance during an airborne trajectory of a person. He stated that speed loss from air resistance could be neglected at projection speeds below 25 mph (40 kph). At projection speeds exceeding 25 mph, on the other hand, he stated that the calculated projection speed should be corrected to account for speed loss due to air drag by adding a correction to the calculated speed. Collins proposed the following table of correction factors for apogee heights of 5 and 7 ft (Table 11.1).

TABLE 11.1 Corrections for initial projection velocity of airborne person

Uncorrected speed (mph)	Correction for 5 ft. apogee (mph)	Correction for 7 ft. apogee (mph)
25	3	3
30	4	5
35	5	6
40	6	8
45	8	10
50	10	13

The assumption of a horizontal launch makes Collins' model too restricted for the analysis of the trajectories of occupants ejected during a rollover. A less restricted model was offered by Searle in 1983 [2] and 1993 [3]. He derived the following formula for determining a pedestrian's projection velocity, based on the conceptual model depicted in Figure 11.1. In this equation, μ is the coefficient of friction between the pedestrian and the ground; g is the acceleration due to gravity; s is the throw distance, including the airborne and sliding/tumbling phases; and θ is the projection angle. This model improved on the Collins model by incorporating a launch angle and also incorporating the ground plane speed loss due to the pedestrian landing on the ground after vaulting through the air. Still, this model needs further developed since a method is needed to relate the projection speed of an ejected occupant to the translational and rotational motion of the rolling vehicle:

$$v_{proj} = \frac{\sqrt{2\mu g s}}{\cos\theta + \mu \sin\theta} \tag{11.2}$$

Aronberg [4] derived another alternative to the Collins model. His model utilized "the time of the body in the air, at the launch velocity, to determine the adjustment necessary to the launch velocity to account for air drag. In addition, the component of velocity corresponding to total horizontal distance traveled was limited to the horizontal velocity component of the body following the airborne trajectory." Aronberg's model is shown graphically in Figure 11.2, along with the meaning of the variables within the model.

The vertical velocity through time is given by the following equation. In this equation, V_y is the vertical velocity, V_0 is the total initial projection velocity, θ is the launch angle, g is the gravitational acceleration, and t is time:

$$V_y = V_0 \sin\theta - gt \tag{11.3}$$

FIGURE 11.1 Searle's model (Figure 5, 1983).

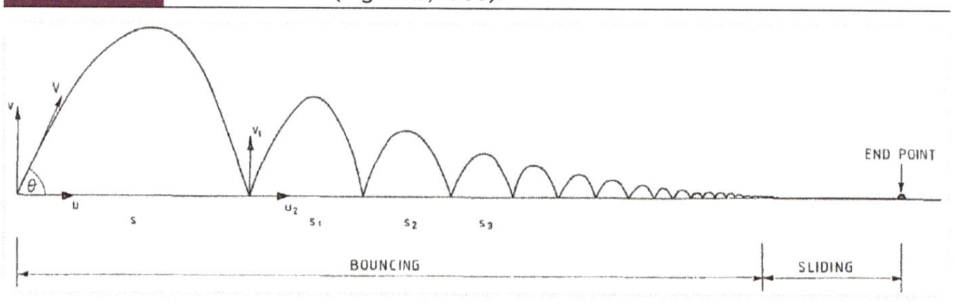

FIGURE 11.2 Aronberg model (1990, Figure 1).

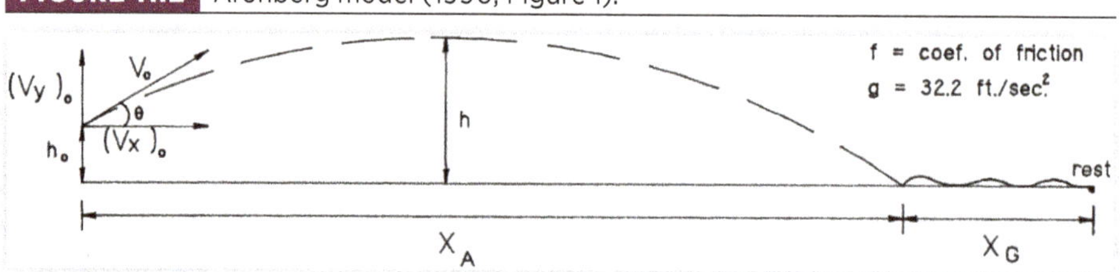

Integrating Equation (11.3) yields the vertical position through time, as follows:

$$y = h_0 + V_0 t \sin\theta - \frac{gt^2}{2} \tag{11.4}$$

At the apogee of the airborne trajectory, $y = h$ and $V_y = 0$. Thus, the elapsed time and trajectory height at the apogee can be calculated with the following equations:

$$t_{apogee} = \frac{V_0 \sin\theta}{g} \tag{11.5}$$

$$h = h_0 + V_0 t_{apogee} \sin\theta - \frac{g t_{apogee}^2}{2} \tag{11.6}$$

When the body strikes the ground, $y = 0$. In this equation, t_{land} is the time at which the body lands on the ground:

$$0 = h_0 + (V_0 \sin\theta) t_{land} - \frac{g}{2} t_{land}^2 \tag{11.7}$$

Applying the quadratic equation to calculate t_{land} yields the following:

$$t_{land} = \frac{V_0 \sin\theta + \sqrt{(V_0 \sin\theta)^2 + 4\frac{g}{2}h_0}}{g} \tag{11.8}$$

The ground plane distance traveled by the projected person during the airborne trajectory can now be calculated with the following equation:

$$x_A = V_0 \cos\theta \cdot t_{land} \tag{11.9}$$

In addition to the airborne distance, the projected person will also typically slide and tumble upon landing. The slide/tumble distance can be calculated with the following equation. In this equation, f is the coefficient of friction (or drag factor) of the pedestrian for the sliding and tumbling phase. Since Aronberg's model does not incorporate a separate term for the speed loss due to landing, the drag factor in Equation (11.10) would be one that incorporates this speed loss:

$$x_G = \frac{(V_0 \cos\theta)^2}{2gf} \tag{11.10}$$

The total throw distance, d_t, is therefore given by the following equation:

$$d_t = x_A + x_B = \frac{V_0^2 \cos\theta \sin\theta + \sqrt{(V_0 \sin\theta)^2 + 4\frac{g}{2}h_0}}{g} + \frac{V_0^2 \cos^2\theta}{2gf} \tag{11.11}$$

Aronberg recognized that his model neglected speed loss due to air resistance. In order to incorporate this speed loss, he examined published data on skydiving free fall velocity with respect to time. Aronberg used this data to produce the graph of Figure 11.3. This graph shows that the speed loss due to air resistance generally increases linearly up to a speed of around 60 mph. At a speed of 61 mph, air resistance would cause just under 3 mph of speed loss for every second the pedestrian is in the air. Above 61 mph, the air resistance continues to increase, but at a greater rate than at speeds below 61 mph. Aronberg notes that "the free fall data presented and air drag derived are for

FIGURE 11.3 Speed loss due to air resistance [4].

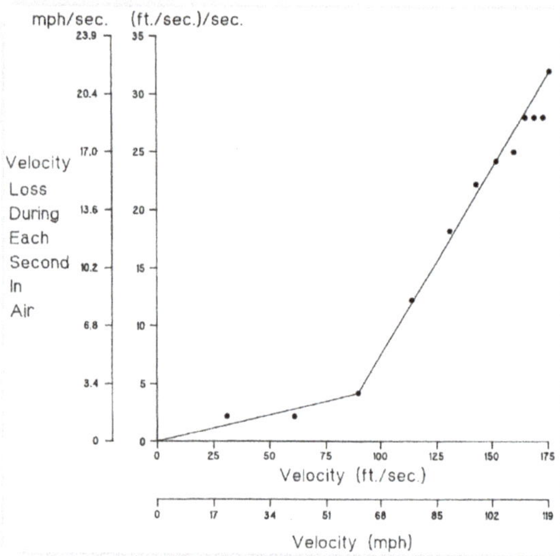

humans in a spread position. Therefore, this graph represents the maximum expected air drag." Aronberg observed that the data he obtained yielded corrections significantly lower than what Collins suggested and that Collins' methods overestimate the speed loss due to air resistance. Like Searle's model, Aronberg's model improves on the Collins model but needs further development before it could be applied to analyze the motion of an ejected occupant - namely a way to relate the projection conditions of the person to the motion of the vehicle.

Funk and Luepke [5] developed a two-dimensional particle model to predict the motion of an occupant ejected during a rollover, including the airborne, landing, and tumbling phases, along with a way to relate the projection conditions of the occupant to the motion of the rolling vehicle. They modeled the occupant as a projectile that was launched at a given radius from the center of gravity of the vehicle, then lands and slides or tumbles. The development of this model follows.

To develop the equations for the airborne phase of an occupant ejection, consider Figure 11.4, which depicts the cross section of a rolling vehicle that is also airborne. The vehicle in Figure 11.4 is depicted with a passenger's side-leading orientation, although for the equations of this model, this does not matter. A coordinate system has been established such that the x-axis runs along the ground and the z-axis is the vertical axis. The origin of this coordinate system is on the ground underneath the vehicle center of gravity at the beginning of the roll phase.

FIGURE 11.4 Airborne portion of the occupant ejection model.

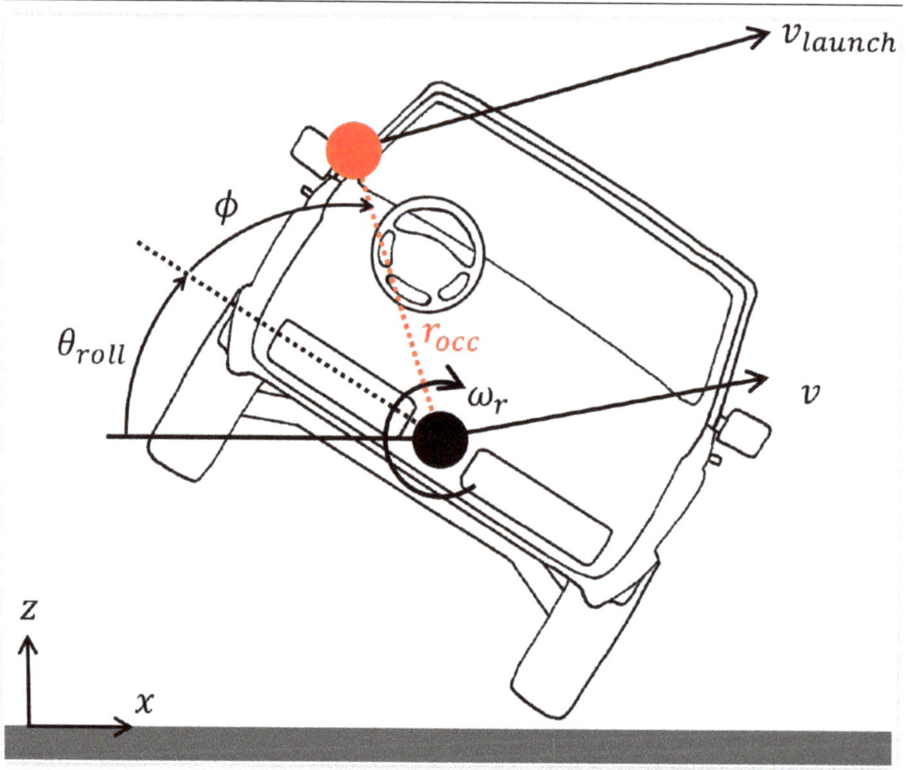

The vehicle depicted in Figure 11.4 has a translational velocity, v, that has both an over-the-ground component and an upward component. It also has a roll angular velocity designated by ω_r. The roll angle of the vehicle is designated by θ_{roll}. The position on the perimeter of the vehicle at which the occupant is ejected is depicted in Figure 11.4 with a red circle. This position is located with a radius measured from the center of gravity, r_{occ}, and with an angle ϕ.

The components of the velocity with which the occupant will be thrown are given by the following equations:

$$v_{x,launch} = v_x + \omega_r \cdot r_{occ} \cdot sin(\theta_{roll} + \phi) \tag{11.12}$$

$$v_{z,launch} = v_z + \omega_r \cdot r_{occ} \cdot cos(\theta_{roll} + \phi) \tag{11.13}$$

Funk and Luepke suggest values for ϕ of 35° for an occupant loaded by the lower portion of the window opening and ejected through the trailing-side window and 130° for an occupant loaded by the roof rail and ejected through the leading-side window.

Based on the vertical velocity of the occupant at the time of the ejection, Equation (11.14) will yield the maximum height attained by the occupant during the airborne phase (h_{max}). In this equation, z_{launch} is the height of the occupant off of the ground at the time of the ejection and g is the gravitational constant:

$$h_{max} = z_{launch} + \frac{v_{z,launch}^2}{2g} \tag{11.14}$$

The height at which the ejection occurs (z_{launch}) can be estimated with the following equation:

$$z_{launch} = r_{occ}\left(1 + \sin(\theta_{roll} + \phi)\right) \tag{11.15}$$

Equation (11.15) assumes the vehicle is not airborne at the time of the ejection, and it treats the vehicle as a rolling cylinder with radius r_{occ}. Alternatively, if the vehicle is airborne and the height of its center of gravity can be determined, the following equation can also be used to obtain the height of the objection:

$$z_{launch} = z_{cg} + r_{occ}\sin(\theta_{roll} + \phi) \tag{11.16}$$

The following equation will yield the position along the ground at which the ejection occurs:

$$x_{launch} = x_{cg} - r_{occ}\cos(\theta_{roll} + \phi) \tag{11.17}$$

Funk and Luepke note that the maximum height (h_{max}) is only attained by an occupant who is ejected with an upward component to their launch velocity. If an occupant is ejected downward, then this height represents the equivalent fall height of the occupant's trajectory. The downward velocity of the ejected occupant can be calculated based on h_{max}, as follows:

$$v_{z,land} = \sqrt{2gh_{max}} \tag{11.18}$$

In this model, air resistance is neglected and the over-the-ground component of the launch velocity is assumed to be equal to the over-the-ground component of the landing velocity:

$$v_{x,launch} = v_{x,land} \tag{11.19}$$

The time the ejected occupant spends airborne is given by the following equation:

$$t_{air} = \frac{v_{z,launch}}{g} + \sqrt{\frac{2h_{max}}{g}} \tag{11.20}$$

And then, the distance the ejected occupant is thrown is given by the following equation:

$$d_{throw} = v_{x,launch} \cdot t_{air} \tag{11.21}$$

The occupant lands at a location given by the following equation:

$$x_{land} = x_{launch} + d_{throw} \tag{11.22}$$

When the occupant lands, a loss in over-the-ground velocity occurs that is given by the following equation:

$$\Delta v_{x,land} = \mu \cdot v_{z,land} \tag{11.23}$$

In this equation, μ is the sliding coefficient of friction between the ground and the ejected occupant. The distance the ejected occupant will slide after landing is given by the following equation:

$$d_{slide} = \frac{\left(v_{x,land} - \Delta v_{x,land}\right)^2}{2\mu g} \tag{11.24}$$

And finally, the position at which the ejected occupant will come to rest is given by the following equation:

$$x_{rest} = x_{land} + d_{slide} \tag{11.25}$$

In applying this model, a reconstructionist will often know where the ejected occupant came to rest. Based on their reconstruction of the rollover phase, including the over-the-ground speed and roll angular velocity histories, they will be determining possible ejection points consistent with the known occupant rest position. A determination of where each possible ejection portal opened during the rollover (where the glass from a particular window broke, for instance) will help to rule out some of the possibilities. Analysis of occupant ejections can also be aided by documenting witness marks on the interior of the vehicle. In some instances, the initial seating position of an occupant will also be known.

Applying the Occupant Ejection Model

To illustrate the application of this occupant ejection model, consider the dolly rollover test involving a 1998 Ford Expedition that Luepke [6] reported in 2007. At the beginning of this test, the Expedition was positioned on the dolly with the driver's side leading and with an initial roll angle of 23° measured from the horizontal. The leading-side tires were positioned against a 4 in. (10.2 cm) flange with the leading tires initially 9 in. (23 cm) above the ground surface. The Expedition exited the dolly at a speed of 43.2 mph (69.7 kph) and then rolled onto a desert earth surface. The Expedition rolled 4 times in approximately 120 ft (37 m).

Hybrid-II 50th-percentile anthropomorphic test devices (ATDs) were initially situated in the Expedition at the outboard seating positions for the first, second, and third rows of the vehicle (positions designated as 1L, 1R, 2L, 2R, 3L, and 3R). Five of the six ATDs were ejected during the rollover (all but 2L), out of the window opening next to which they were initially situated. Funk and Luepke noted that "the time of occupant ejection could only be determined approximately from the video, since the entire ejection process occurred over approximately a quarter of a second and involved ATD interaction with a variety of vehicle structures. The exact time of ejection was chosen based on the 'slingshot release' assumption so that the predicted trajectory of the dummy matched the observed trajectory. At this time point, the ATD was approximately halfway out of the window opening. It was possible to determine the location of landing for all of the ejected ATDs from photographs of disturbances in the ground. The time of landing was defined as the moment when the center of gravity of the dummy struck the ground in the video."[11]

Figure 11.5 is a scale diagram published by Luepke that shows the vehicle motion in quarter roll intervals along with other physical evidence from the rollover, including the locations at which ATDs came to rest. We have added red rectangles around the text to

FIGURE 11.5 Scale diagram for expedition test, including rest positions of ATDs.

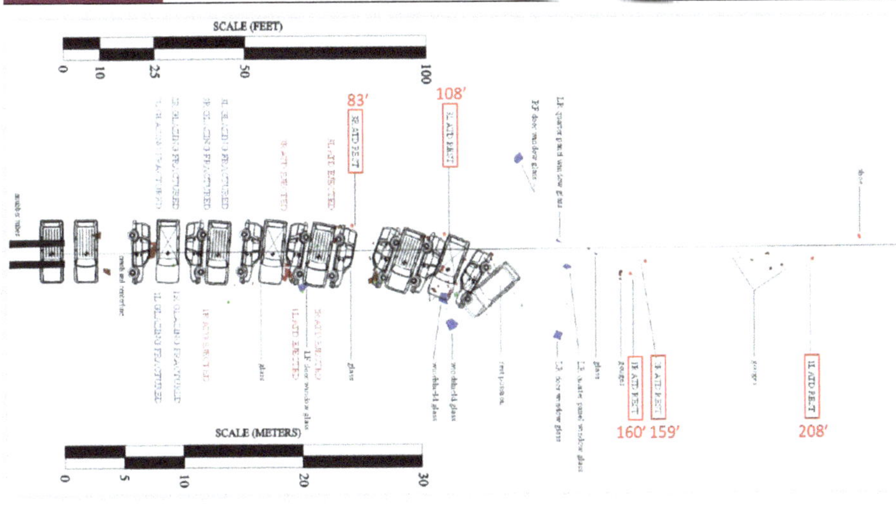

[11] Searle [2] noted that "in most accidents however the point of first landing cannot be determined and it is the point at which the object comes to rest which is recorded." This is certainly true for many real-world rollovers. If the analyst can determine an occupant's landing position, then this position can be used in the analysis. If the landing position cannot be determined, the analysis can still proceed without it. The analyst will have to find a different criteria with which to rule out certain possible occupant ejection paths. [See also, the Searle article from 1993 [3].]

highlight the positions at which the 5 ejected ATDs came to rest. In addition, we have added a dimension to each ATD rest position indicating the distance at which the ATD came to rest when referenced off of the vehicle position at its exit from the dolly. Funk and Luepke reported rest positions from the end of the first tire marks of the test; thus, there is a 17-foot discrepancy between the rest position dimensions we report here and those they reported.

To illustrate the application of the occupant ejection model, consider the ATD that came to rest closest to the initiation of the roll: the right rear ATD (3R) came to rest approximately 83 ft from where the vehicle exited the dolly. In analyzing the ejection of this ADT, the actual over-the-ground and roll angular velocities of the vehicle were utilized (see Figures 11.6 and 11.7). The ejection radius was varied between 41 and 47 in., and the ATD to ground coefficient of friction was varied between 0.5 and 0.7.

Mathematically, there were 5 possible vehicle positions along the roll path where this ejection of ATD 3R could have occurred. The *first* of these was when the vehicle had traveled around 21 ft and rolled approximately 115° to 135°. In reality, the ejection could not have occurred at this position because the passenger-side roof of the vehicle had not struck the ground and the window glass adjacent to this ATD was still intact. Thus, this possibility could be ruled out. The *second* possible location was when the vehicle had traveled around 37 ft and rolled approximately 295° (0.82 rolls). The glazing separated from the vehicle at the 3R position around this time period, and thus, it was possible that the ejection could have taken place at this point in the rollover. For the midrange assumptions on the radius and coefficient of friction, the mathematical model, in this instance, predicted a throw distance of approximately 51 ft and a sliding distance of approximately 1 ft. The third possible location was when the vehicle had traveled approximately 52 to 53 ft and rolled approximately 480° to 498° (1.35 to 1.38 rolls). This scenario corresponds closely to the distance at which the ATD was actually ejected (a vehicle position of 54 ft). For the midrange assumptions on the radius and coefficient of friction, the mathematical model, in this instance, predicted a throw distance of approximately 7 ft and a sliding distance of approximately 21 ft. The actual throw distance for this ATD was 16 ft and the sliding distance was 11 ft. The fourth possible location was when the vehicle had traveled approximately 65 ft and rolled approximately 650° (1.81 rolls). For the midrange

FIGURE 11.6 Velocity time history for the expedition.

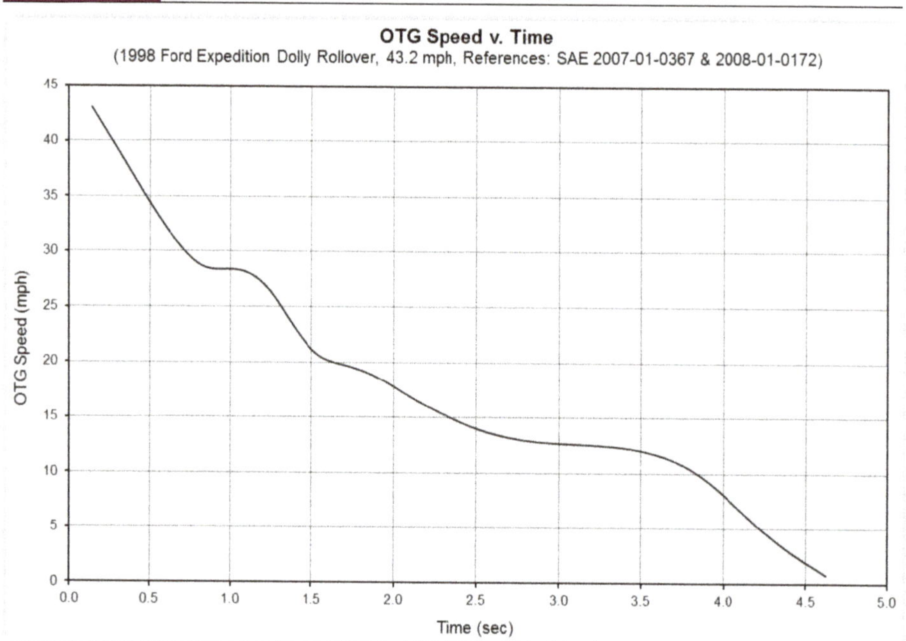

FIGURE 11.7 Roll angular velocity history for the expedition.

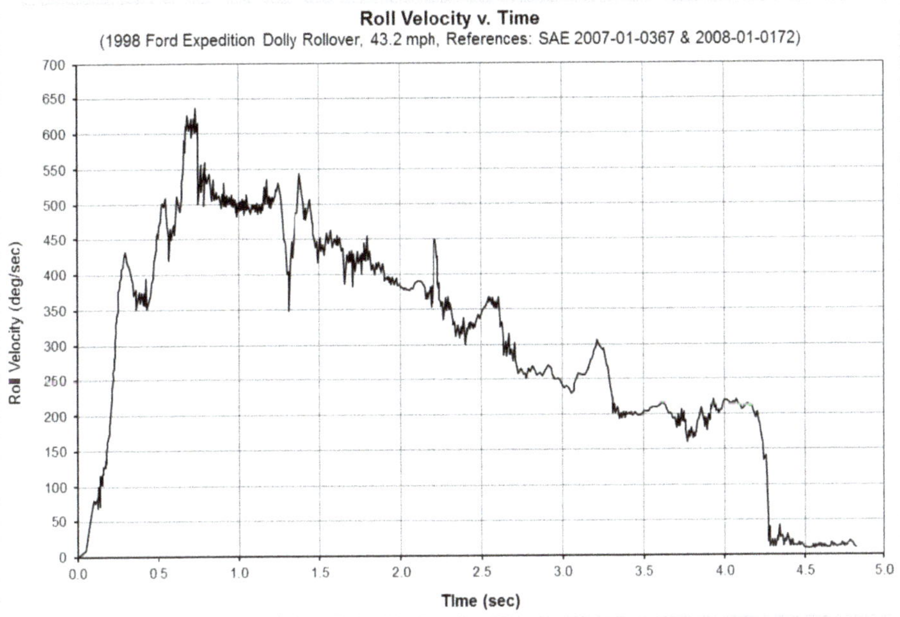

assumptions on the radius and coefficient of friction, the mathematical model, in this instance, predicted a throw distance of approximately 22 ft and zero sliding distance. The fifth possible location was when the vehicle had traveled approximately 85 ft and rolled approximately 958° (2.66 rolls). This final possibility is associated with zero throw and slide distance for the ATD and thus represents the ATD being simply deposited on the ground.

In this instance, there were four possible ejection scenarios that occurred after the glass separated from the window adjacent to 3R, so the question arises as to how, in practice, one would distinguish between these scenarios, without reviewing the test video. Funk and Luepke noted that a similar situation existed for most of the ejections they analyzed: "Most models predicted several possible points of ejection at various stages of the rollover sequence that matched the rest position of the dummy. Of course, only one of these predicted ejection scenarios can be accurate." Thus, additional criteria are needed to determine which of the possible ejection scenarios is the actual one. If, on any particular case, the landing position of an occupant can be determined, in addition to the occupant's rest position, this landing position can be used to distinguish one scenario from another. For the case under consideration here, each possible ejection scenario produced a different throw and slide distance even though the final rest position of the ATD for each scenario was the same.

Another possible criteria is to consider where an occupant comes to rest relative to the rest position of the vehicle. In regards to the test under consideration here, Funk and Luepke note that "three of the ATDs were ejected on the 'high' side, meaning that they left the vehicle with an initially upward trajectory (occupants 1L, 1R, and 2R). All of these ATDs landed and slid to rest well beyond the vehicle's final point of rest. Two of the ATDs were ejected on the 'low' side, meaning that they left the vehicle with an initially downward trajectory (occupants 3L and 3R). Both of these ATDs were initially thrown ahead of the vehicle and then rolled over by the vehicle. Dummy 3L was pushed forward when its back was struck by the top left rear corner of the vehicle, which then rolled over it. Dummy 3R was actually hit twice by the rolling vehicle. After its initial bounce off the ground, it was hit in the face by the left rear bumper. Then the vehicle rolled over its pelvis, leaving an indentation in the left roof rail between the C and D pillars." Thus, the two ATDs that

TABLE 11.2 Vertical velocity directions for possible ejection scenarios

Possible ejection scenario	Direction of vertical launch velocity
1	Downward
2	Upward
3	Downward
4	Upward
5	Level (Slight Upward)

experienced a low side ejection came to rest short of the vehicle rest position and the three that experienced high side ejections came to rest beyond the vehicle rest position.

For the test under consideration here, ATD 3R came to rest short of the vehicle, making a downward launch velocity likely. Table 11.2 lists the direction of the vertical velocity for each of the possible ejection scenarios. Scenario 1 has already been ruled out and Scenario 3 is the only other one that has a downward vertical launch velocity.

The above description by Funk and Luepke also notes that ATD 3R was rolled over by the vehicle. This is another criteria that could be used to distinguish between scenarios, since such an occurrence would likely be evidence from an ejected occupant's injuries. Hovey, Kaplan, and Piziali affirm this type of rationale, noting that "an ejectee presenting with extensive road rash type abrasions may indicate an ejection solution with a large slide distance is more likely than an ejection solution with a small slide distance. An ejectee presenting with crush injuries may indicate an ejection solution where there is post-ejection contact, e.g., the ejectee lands ahead of the vehicle and the vehicle overruns the ejectee." Finally, Funk and Luepke noted that "in all of the ejections, there was a forward component to the tangential velocity vector. In other words, all of the ejections occurred at a release angle ($\theta + \varphi$) between 0° and 180°. There were no 'deposits' in which the dummy was ejected at a low speed near the ground. All of the dummies experienced high energy impacts upon landing…"

Another criterion for distinguishing between possible ejection scenarios has been developed by Hovey et al. [7, 8]. The noted that the yaw angle and yaw rate of the vehicle will influence the direction in which the occupant is ejected and the ultimate resting position of the occupant. In their model, an ejection scenario could produce the right resting position for the occupant in terms of the distance along the roll path, but if it does not also produce the correct position lateral to the roll path, it can be ruled out.

References

1. Collins, J.C., *Accident Reconstruction*, (Springfield, IL: Charles C Thomas Publisher, 1979), ISBN 0-398-03907-0.
2. Searle, J. and Searle, A., "The Trajectories of Pedestrians, Motorcycles, Motorcyclists, etc., Following a Road Accident," SAE Technical Paper 831622, 1983, doi:10.4271/831622.
3. Searle, J., "The Physics of Throw Distance in Accident Reconstruction," SAE Technical Paper 930659, 1993, doi:10.4271/930659.
4. Aronberg, R., "Airborne Trajectory Analysis Derivation for Use in Accident Reconstruction," SAE Technical Paper 900367, 1990, doi:https://doi.org/10.4271/900367.
5. Funk, J. and Luepke, P., "Trajectory Model of Occupants Ejected in Rollover Crashes," SAE Technical Paper 2007-01-0742, 2007, doi:10.4271/2007-01-0742.
6. Luepke, P., Carhart, M., Croteau, J., Morrison, R. et al., "An Evaluation of Laminated Side Window Glass Performance During Rollover," SAE Technical Paper 2007-01-0367, 2007, doi:10.4271/2007-01-0367.
7. Hovey, C., Kaplan, M., and Piziali, R., "Occupant Trajectory Model Using Case-Specific Accident Reconstruction Data for Vehicle Position, Roll, and Yaw," *SAE Int. J. Passeng. Cars - Mech. Syst.* 1, no. 1 (2009): 396-405, doi:10.4271/2008-01-0517.
8. Hovey, C., Raphael, E., and Xu, H., "Validation of Occupant Trajectory Model Using the Ford Expedition Dolly Rollover Experimental Test Data," SAE Technical Paper 2013-01-0472, 2013, doi:10.4271/2013-01-0472.

about the authors

Nathan Rose is a Director and Principal Crash Reconstructionist at Kineticorp, LLC, a Denver-based crash reconstruction, forensic engineering, and forensic visualization firm that he helped found in 2005. Prior to that, he held positions as an engineer (1998 to 2003) and a senior engineer (2003 to 2005) at Knott Laboratory, another Denver-based forensic engineering firm. He holds a bachelor's degree in Engineering from the Colorado School of Mines (1998) and a master's degree in Mechanical Engineering from the University of Colorado at Denver (2003). Nathan is accredited as a Traffic Accident Reconstructionist by the Accreditation Commission for Traffic Accident Reconstruction (ACTAR) and he has offered expert testimony as a reconstructionist in courts around the United States. During his graduate studies, he specialized in dynamics and impact mechanics and he has published numerous technical articles and reports related to vehicular crash reconstruction. These articles have covered topics including crush analysis, structural restitution, video analysis, crash test sensor analysis, rollover crash reconstruction and analysis, and motorcycle crash dynamics and reconstruction. Nathan is a past organizer for the Rollover, Rear Impact, and Accident Reconstruction Sessions held annually at the SAE World Congress. He teaches a three day accident reconstruction course through SAE

Gray Beauchamp is a Principal Engineer at Kineticorp, LLC and is an expert in traffic accident reconstruction and vehicle dynamics. He holds a Bachelor's degree and Master's degree in Mechanical Engineering and is a Professional Engineer in the state of Colorado. Since 2003, he has investigated and reconstructed hundreds of vehicular accidents involving passenger vehicles, commercial vehicles, motorcycles, bicycles and pedestrians. He has also reconstructed accidents involving recreational vehicles. Gray has extensive experience as an expert witness in court and deposition. Gray has authored numerous peer-reviewed publications, many of which have focused on single vehicle loss of control accidents. Specifically, he has researched vehicle handling during tire disablements, the interpretation of tire mark evidence, and the various phases of rollover accidents. Much of his research has included full scale vehicle testing. He is a peer reviewer for technical papers published in the Rollover Session, the Accident Reconstruction Session, and in the Analysis, Synthesis, and Design of Advanced Suspensions Session of SAE. Gray is an instructor for Reconstruction and Analysis of Rollover Crashes of Light Vehicles, a seminar offered through SAE. He guest lectures on Accident Reconstruction at the Denver Police Department for officers training in the Level 3 Advanced Traffic Collision Investigation and Reconstruction Course. He has also guest lectured at the graduate level at the University of Colorado and Wayne State University.

Alan F. Asay is a principal engineer and reconstructionist at Asay Engineering LLC. He holds both a bachelor's and a master's degree in mechanical engineering from Brigham Young University located in Provo, Utah. He is a licensed professional mechanical engineer in Utah and has over thirty years of experience in investigation, consulting, testing, and litigation support for clients across the United States. He has been an active member of the Society of Automotive Engineers (SAE) for over twenty-five years, serving as a peer reviewer of technical papers, and has additionally served for the last five years as an organizer of the Accident Reconstruction Session of the annual SAE World Congress. He has published over twelve papers associated with his testing and research in the field of accident reconstruction. He directed and published a number of the full-scale, steering-induced rollover crash tests analyzed in this book.

www.ingramcontent.com/pod-product-compliance
Lightning Source LLC
LaVergne TN
LVHW071341210926
848486LV00003B/205
* 9 7 8 0 7 6 8 0 9 3 7 2 8 *